OCCUPATIONAL HYGIENE MANAGEMENT GUIDE

Stanley E. Jones
H. Roland Hosein
Glenn R. Swalm
Jack F. Yablonsky

LEWIS PUBLISHERS

Library of Congress Cataloging-in-Publication Data

Occupational hygiene management guide/edited by Stanley E. Jones . . . [et al.].

p. cm.
Includes bibliographical references.
1. Industrial hygiene. I. Jones, Stanley E.
HD7261.034 1990 658.3'82—dc20 89-14532
ISBN 0-87371-255-2

LEWIS PUBLISHERS, INC.
121 South Main Street, Chelsea, Michigan 48118

PRINTED IN THE UNITED STATES OF AMERICA

Preface

The development and application of occupational hygiene principles to protect employee health has been an evolving discipline for over 2000 years. History records that many scholars down through the centuries, including Hippocrates, have written about diseases resulting from occupational exposures to such chemicals as zinc, sulfur, lead, mercury, and carbon monoxide. These chemicals and many others, along with certain physical hazards (e.g., radiation), are now a necessary part of the day-to-day production operations of many industrial facilities.

The scientifically recognized and documented harmful effects of these toxic and physical agents confer an obligation on each place of business to operate an effective preventive employee health program. This program must recognize and address not only the short-term effects of many of these agents but also the long-term effects of materials such as asbestos, exposure to which can result in asbestosis some 40 years later. Failure to pay sufficient attention to these kinds of issues can have catastrophic results, not only for individuals (e.g., workers exposed to vinyl chloride or lead) but also for a business (e.g., Johns Manville Co., which was forced into bankruptcy and Chapter 11 reorganization by asbestos litigation).

Successful management of these types of issues requires that strong, effective programs be in place at all times. The *Occupational Hygiene Management Guide* has therefore been prepared to provide guidelines and recommendations that enable local management to develop and implement effective programs. These efforts minimize the potential for personal as well as business liabilities that can ultimately result from exposing employees to environmental agents with adverse health effects. Hazard communication regulations and media coverage of occupational health issues, as well as some publicized criminal proceedings against employers who have ignored occupational health concerns, have heightened public awareness of the need for everyone to become more involved.

Some diseases caused by exposure to toxic substances take upwards of 40 years after exposure to manifest. Other chemicals or physical agents can cause acute health effects immediately after exposure. This demonstrates that protecting the worker must start the first day on the job and continue throughout his or her working career. For a program to be totally effective, the design, implementation, and administration of occupational hygiene programs need the input, direction, and personal involvement of qualified professionals and the personal involvement of all levels of management, other professionals, and the employees on the factory floor and in service and office environments.

The purpose of this book is to provide a reference for individuals who have a direct or indirect responsibility for employee welfare. Written to be applicable worldwide, it is designed to provide upper and middle management with a better understanding of the subject of occupational hygiene; to give the local management (e.g., plant managers, employee relations managers, manufacturing managers, and engineers) the knowledge necessary to design, implement, and administer a program; and finally to outline the specific roles of each member of the management team.

The book is divided into four sections. The first provides a program overview with an executive summary. It also describes what the book is designed to do. This section discusses allocation of occupational health responsibilities, detailing who in management or in operations is responsible for ensuring that the various elements of an occupational hygiene program are given the attention they require.

The second section discusses and describes hazard recognition, evaluation, and control procedures. The practices involved are (a) recognition, an assessment of the workplace environment to identify potentially hazardous exposures; (b) evaluation, the scientific monitoring process to qualify and quantify the exposures; and (c) control, the determination of the systems necessary to minimize or remove potential exposures. It also highlights types of programs and procedures that must be in place for a program to be totally effective.

Section 3 details the auditing and measurement process and outlines the procedures necessary to monitor the actual effectiveness of the total program, both short- and long-term.

The final section is a set of six appendices. Appendix A lists typical program elements. Appendix B discusses some typical chemical and physical agents that can be found in the work environment, the possible sources of these agents, and their potential health effects. Appendix C summarizes the principles of toxicology and their relationship to the field of occupational hygiene, using tables to show certain acute and chronic harmful effects resulting from overexposure to selected chemical and physical agents.

Appendix D contains the 1989–1990 Threshold Limit Values (TLVs) and Biological Exposure Indices (BEIs) published by the American Conference of Governmental Industrial Hygienists (ACGIH). Appendix E provides a list of recommended reference materials, and Appendix F is a glossary of terms used by health professionals.

Acknowledgments

The authors would like to thank the following people who generously and extensively gave of their expertise and time to assist us with the design and content of this book: Toni Federowski, PhD, DABFT, who provided guidance with the toxicology references, and Dan Uyesato, Esq., who provided legal review and input.

Stanley E. Jones is currently Corporate Program Manager of Industrial Hygiene for General Electric Company in Fairfield, Connecticut. He is responsible for monitoring and supporting the development and operation of effective occupational hygiene programs throughout General Electric's worldwide operations and providing technical assistance to company management and health professionals through the processes of review, consultation, recommendation, liaison and representation, and program appraisal. Born and educated in England, Mr. Jones majored in electrical engineering at a local technical college. He began his career with Imperial Chemical Industries, with responsibilities allied to the maintenance, construction, and commissioning of chemical and petrochemical plants. After coming to the United States in 1965, he accepted a position with The Budd Company in its Detroit Automotive Division. Since joining General Electric Company in 1966, his positions have included Manager of Health and Safety Assurance for General Electric's Engineered Materials Group (six specialty chemical businesses); Manager of Health, Safety and Environmental Engineering; and Specialist in Safety and Environmental Engineering for the Metallurgical Division. In these positions, Mr. Jones was responsible for the design, implementation, and overall effectiveness of industrial hygiene, safety, and environmental protection programs and regulatory compliance for as many as 35 U.S. and worldwide manufacturing operations.

Mr. Jones is certified by the American Board of Safety Professionals and the Hazard Control Manager Certification Board and is currently pursuing certification by the American Board of Industrial Hygiene. He is a member of the American Industrial Hygiene Association (AIHA), the American Society of Safety Engineers (ASSE), the American Public Health Association (APHA), MEDICHEM, and the National Safety Management Society (NSMS). He is a founding member of the NSMS Michigan Chapter and was its vice president prior to relocating to Connecticut in 1985. Mr. Jones was elected to the Board of Trustees of the Industrial Health Foundation, Inc., in 1988 and to the Governing Board of NSMS in 1989. He was recently appointed to the Health and Safety Steering Committee of the Organization of Resource Counselors, Inc. (ORC), a trade association located in Washington, D.C.

Mr. Jones has presented a number of papers, at national and international meetings, discussing the professional's management role, effective health and safety programs, the similarities and differences of international health and safety regulations and programs, and auditing in the international arena. During the past 18 months, he has been involved with the dean and faculty of Harvard University's School of Public Health in the successful development and promotion of an MPH program that provides for students to spend their third semester with a major industry.

H. Roland Hosein is currently Manager of Corporate Health, Safety, and Environment for General Electric Canada, Inc. In this role Dr. Hosein has developed and implemented policies and practices consistent with the company's objectives and with regulatory requirements. Additionally, he is the senior specialist advisor to the company's Canadian and South and Central American operating divisions and plants.

Prior to joining General Electric Canada in 1980, he was Section Head, Chemical Hazards and Toxicology, Alberta Department of Labour; Research Associate, Lung Research Center, Yale University Medical School; and Director of Industrial Hygiene, Faculty of Medicine, University of Toronto.

In addition to his role at GE Canada, Dr. Hosein is Associate Professor at the University of Toronto, Faculty of Medicine, Department of Occupational and Environmental Health. He is author of over 30 technical papers and numerous monographs and modules.

He received his BSc in chemistry and MSc in industrial hygiene from London University and his PhD in environmental epidemiology from the University of Western Ontario. He is certified by the American Board of Industrial Hygienists and registered by the Canadian Registration Board of Occupational Hygiene.

Dr. Hosein holds memberships in AIHA, the American Conference of Governmental Industrial Hygienists (ACGIH), the American Academy of Industrial Hygienists, and the Canadian Public Health Association.

Glenn R. Swalm is currently Manager of Environment, Health and Safety for General Electric's Steam Turbine business in Schenectady, New York. He is responsible for the overall development, implementation, and effectiveness of industrial hygiene, occupational safety, and environmental protection programs and procedures in the manufacturing operations. Previously he was Manager of Industrial Hygiene for the GE Aircraft Engines business, where he was responsible for the design and success of many occupational hygiene programs. Prior to joining General Electric in 1985, Mr. Swalm held a number of industrial hygiene and safety positions with Armco, Inc., including Manager of Safety and Industrial Hygiene for Armco's specialty steel business; Corporate Industrial Hygienist servicing its steel, coal, and oil field equipment operations; and Industrial Hygienist in its steel business.

He received his BSc in physics and MSc in industrial hygiene from Central Missouri State University. He is certified by the American Board of Industrial Hygiene and the American Board of Safety Professionals. Mr. Swalm is a member of AIHA and the American Academy of Industrial Hygiene. He has held a number of board positions in his local AIHA chapter.

Jack F. Yablonsky is currently the Industrial Hygiene Manager for the General Electric Plastics Group, headquartered in Pittsfield, Massachusetts, with resin, compounding, and specialty chemical manufacturing operations located worldwide. This position, held since 1984, includes responsibility for development and oversight of businesswide industrial hygiene programs, as well as management of industrial hygiene and safety programs for the headquarters R&D site. Mr. Yablonsky previously held the position of Senior Industrial Hygienist for the General Electric Turbine business group. Prior to joining GE, he held investigative and supervisory positions with the Occupational Safety and Health Administration.

Mr. Yablonsky received a BSc in biology from the University of Cincinnati (1973), an MSc in environmental health (industrial hygiene) from the University of Cincinnati College of Medicine (1975), and an MBA from Rensselaer Polytechnic Institute (1984). He received American Board of Industrial Hygiene certification in 1980 and is a board member and past president of the upper New York State AIHA chapter.

Contents

Section 1: Program Overview

Executive Summary 1
Purpose 2
Scope 2
Responsibility 3
Management Function Responsibility Chart 4
Sample Company Policy 6
Sample Procedure 9

Section 2: Program Guidelines and Recommended Practices

Introduction 13
Administration 14
Recognition 16
Exposure Assessments 18
Hazard Reduction Procedures 21
Hazard Communication 29
Training 30
Recordkeeping 31
Use of Support Laboratories/Consultants 32

Section 3: Auditing and Measurement

Introduction 35
Auditing and Measurement Procedures 35
Organization and Procedures 36
Occupational Illness and Injury Experience 36
Hazard Identification and Control 37
Training and Communications 38

APPENDICES

Appendix A: Typical Program Elements 41

Appendix B: Chemical and Physical Agents: Descriptions and Sources 43

Appendix C: Toxicology 47

Appendix D: Threshold Limit Values and Biological Exposure Indices for 1989–1990 55

Appendix E: Recommended Reference Materials..... 193

Appendix F: Glossary 197

OCCUPATIONAL HYGIENE MANAGEMENT GUIDE

1

Program Overview

EXECUTIVE SUMMARY

Audit programs and other onsite management and professional reviews point out that occupational hygiene programs in business operations are not always adequate in quality and content. To modify this situation, managers and engineering professionals must be an integral part of a successful program and must become familiar with the elements that comprise it. This book is designed to assist all levels of management in more fully understanding typical program elements and their personal roles in the design, implementation, and administration of such programs.

Occupational illnesses can be caused by *chemical agents,* such as dusts, fumes, mists, vapors, and gases, or by *physical agents,* such as noise, temperature extremes, and radiation. Employees may suffer adverse health effects when overexposed to these agents, i.e., when subjected to an exposure in excess of recognized safe limits. Effects may develop after repeated exposures over long periods of time (*chronic*) or through a single exposure event (*acute*). These illnesses or diseases can cause temporary or permanent disability, impact the quality of life, or result in a shortened life span. Effective occupational hygiene programs minimize such risks.

Managers and supervisors play a key role in preventing occupational illnesses. Management knowledge of products, processes, workplace hazards, and exposure potentials is essential to the development of effective programs and controls designed to minimize employee exposures. Furthermore, all levels of management should be assigned responsibilities for protecting the health and safety of employees according to company policy, which should be holistic in design and should include guidelines on health, safety, and environmental protection. This is because occupational hygiene can affect and be affected by other environmental and workplace issues and concerns. (A **Sample Company Policy** is included in this section.)

While the primary objective of a comprehensive occupational health program is to ensure the continued protection of the health of all employees, such programs do result in other tangible benefits, including control of workers' compensation costs, reductions in lost time, and minimization of future personal and business liability (e.g., toxic torts, criminal lawsuits).

For a company to be competitive, it is essential that business locations provide a safe and healthy working environment for one of their most important resources: their employees. This objective should be consistent with a corporate commitment to leadership in all business endeavors.

PURPOSE

The purpose of this book is to provide management with guidelines and recommended practices for maintaining present occupational hygiene programs and for designing and implementing new programs where needed to protect employee health.

SCOPE

This management guide provides the framework of an occupational hygiene program and is designed to be applicable worldwide, in any type of service, commercial, or industrial setting.

RESPONSIBILITY

Primary responsibility for protecting the health of employees must be assigned to each operating manager. The specifics of this responsibility should be detailed in a company policy. The policy should specify that *each manager with overall responsibility and authority for operating a facility or activity or for offering a product or service is responsible for protecting employee health.* The policy should further state that part of that responsibility is to ensure *the development and implementation of effective practices and programs for health, safety, and environmental protection.* (See the Management Function Responsibility Chart for suggested position-specific responsibilities.)

Table 1.1 Management Function Responsibility Chart

	President/ CEO	Vice President	General Manager	Manager, Occupational Hygiene	Manager, Occupational Safety	Medical Director/ Nurse	Manager, Environmental Protection
Program Management							
Communicate corporate policy	X	X	X				
Establish & communicate program objectives	X	X	X				
Provide motivation & program support		X	X	X	X	X	
Monitor program progress & effectiveness	X	X	X	X	X	X	
Ensure overall program development and management		X	X				
Hazard Recognition							
Evaluate/survey process & work options				X	X	X	X
Evaluate/survey materials & equipment				X	X	X	X
Identify illness or disease patterns				X		X	
Communicate new process or material information				X	X		X
Obtain appropriate MSDS				X	X	X	X
Review appropriate MSDS				X	X	X	X
Hazard Assessment							
Conduct exposure evaluation/employee monitoring				X			
Identify all exposure routes (process observation)				X		X	
Provide medical surveillance & diagnosis						X	
Seek professional occupational hygiene services					X	X	
Hazard Control							
Design/install/maintain equipment for hazard reduction				X	X		X
Design & monitor process for hazard reduction				X			X
Insure purchase of approved equipment and material				X	X	X	X
Inspect and/or maintain control devices				X			X
Establish/enforce/utilize good work practices							
Develop and/or install control devices							
Regulatory Compliance							
Adhere to appropriate governmental regulations		X	X	X	X	X	X
Interpretation of governmental regulations				X	X	X	X
Maintain necessary records				X	X	X	X
Employee Training							
Ensure development & support of training program		X	X				
Develop required programs				X	X	X	X
Conduct required training programs							
Ensure participation & involvement		X	X				

Manager, Employee Relations	Manager, Production	Line Manager	Production Supervisor	Manager, Manufacturing Engineering	Manager, Facilities or Plant Engineering	Manufacturing Engineer	Maintenance & Service Supervisor	Purchasing Agent	Hourly Employee
X	X								
X	X	X	X				X		
X	X								
	X								
		X	X	X	X	X	X		X
		X	X	X	X	X	X		
	X	X	X	X	X	X	X		
					X	X		X	
				X					
X	X			X					
	X		X	X	X	X	X		
				X	X	X			
	X	X	X	X	X	X	X	X	
		X	X		X		X		
X	X	X	X	X			X		X
				X	X	X	X		
X	X	X	X	X	X	X	X		X
X									
X		X		X	X				
X	X								
X									
X		X	X				X		
X	X	X	X				X		

SAMPLE COMPANY POLICY

XYZ COMPANY

COMPANY POLICY GUIDE

PAGE: 1 of 3

POLICY

Issued by: Corporate Executive Office

TAB: Company Policy Statements

POLICY NO.:

HEALTH, SAFETY, AND THE ENVIRONMENT

NEED FOR POLICY

It is the objective of the XYZ Company to manage its plants and activities so that employees' health and safety on the job are protected from unreasonable risks, so that reasonable employee expectations concerning the work environment are met, and so that the public and the environment are properly protected from adverse effects of plant operation and from use of Company products and services.

A consistent holistic approach is needed in the Company to meet evolving requirements for health, safety, and protection of the environment. A Company Policy is needed to reaffirm XYZ Company's long-standing commitment to high standards of health, safety, and environmental protection, and to assign responsibility and authority for the particular actions and decisions required to achieve the intent of this commitment.

STATEMENT OF POLICY

It is the policy of the XYZ Company:

- to take all appropriate measures to protect the health and safety of its employees in the performance of their assigned work, giving full regard to evolving industry practices, regulatory requirements, and societal standards of care
- to eliminate where possible or limit to lowest practicable levels adverse effects of human health and the environment from its products, services, facilities, and activities

RESPONSIBILITY AND AUTHORITY

Each manager with overall responsibility and authority for operating a facility or activity, or for offering a product or service, is responsible for applying this Policy to the appropriate facility, activity, product, or service

HEALTH, SAFETY, AND THE ENVIRONMENT	**PAGE:** 2 of 3 **POLICY NO.:**

and for assuring development and implementation of effective practices and programs for health, safety, and environmental protection.

In order to assure high standards in medical advice and counsel, all physicians employed or otherwise engaged by the Company are considered *professionally* accountable to the Company Medical Director.

REPORTS AND PERIODIC REVIEWS

At least annually, each Senior Executive, Vice President, Managing Director, or Plant Manager will review, with direct reports and such other employees as the manager may select, the component operations pertaining to all matters the manager believes may present issues under this Policy. The Senior Executive, Vice President, Managing Director, or Plant Manager will annually report the appraisal of any such issues to the cognizant member of the Corporate Executive Office.

RESPONSIBILITY FOR COUNSELING

Corporate Medical will:

- develop and administer objectives and standards for Company-supplied health services
- provide for professional leadership to Company health professionals
- issue appropriate medical bulletins, procedures, and guidelines
- have the authority to conduct appraisals of employee health programs and make appropriate recommendations for actions to be taken by the responsible manager

Corporate Health, Safety, and Environment Operations (CHSEO) will:

- develop and administer objectives for employee health and safety and environmental protection programs.
- provide leadership to Company professionals involved with employee health and safety, occupational hygiene, environmental protection, and hazardous materials transportation
- issue appropriate bulletins, procedures and guidelines.
- have the authority to conduct appraisals of programs addressing occupational hygiene, safety, environmental protection and hazardous materials transportation
- provide functional advice and counsel on health, safety, and environmental protection matters concerning Company facilities, products, and services

HEALTH, SAFETY, AND THE ENVIRONMENT	PAGE: 3 of 3 POLICY NO.:

Corporate Engineering Services (CES) will:

- provide advice and counsel on health and safety matters associated with Company products and services
- issue bulletins and guidelines, as appropriate

Procedure

Please refer to the *Procedure for Implementation of Health, Safety and Environment Policy* issued by the Corporate components responsible for counseling.

SAMPLE PROCEDURE FOR IMPLEMENTATION OF HEALTH, SAFETY, AND ENVIRONMENT POLICY

COMPANY POLICY

Company Policy on health, safety, and the environment commits the Company to take all appropriate measures to protect the health and safety of its employees on the job, as well as the public and the environment. Each manager with overall responsibility and authority for operating a facility or activity or for a product or service offered is responsible for applying this Policy and for assuring that effective practices and programs for health, safety, and environmental protection are developed and implemented.

OPERATING MANAGEMENT ROLE

- Operating management has responsibility for carrying out this Policy and for assuring that medical, health, safety, and environmental protection resources, practices, and programs appropriate to each component's requirements are in place and implemented.
- A comprehensive self-appraisal and planning program shall be utilized by operating management to provide the necessary assurances that employee health, safety, and environmental protection are provided.
- Operating management shall assure that all manufacturing and service facilities and other related locations have access to and receive guidance from physicians and other related professionals.
- Operating management shall establish appropriate procedures and clearly assigned responsibilities to assure that the products they design, produce, sell, and install, and/or the services they provide, eliminate where possible or limit to lowest practicable levels adverse effects on human health and the environment.

CORPORATE HEALTH, SAFETY, AND ENVIRONMENT OPERATIONS ROLE

Corporate Health, Safety, and Environment Operations (CHSEO) will develop, administer, and update objectives, programs, and standards for employee health and safety, and environmental protection, as well as provide leadership and guidance on occupational hygiene and employee safety to associated Company professionals.

In addition to direct consultation as may be requested by operating components, CHSEO will communicate with appropriate Company personnel using information letters and guidelines regarding practices, programs, regulations, and standards.

Corporate staff has the authority to review component facilities and to make recommendations to operating management concerning compliance with Company and regulatory requirements, procedures and practices.

PROCEDURE FOR IMPLEMENTATION OF HEALTH, SAFETY, AND ENVIRONMENT POLICY	**PAGE:** 2 of 3

CORPORATE MEDICAL ROLE

Corporate Medical (CM) will develop, administer, and update objectives and standards for Company health services and will provide functional leadership and guidance to Company health professionals.

CM has the authority to review component facilities and to make recommendations to operating management.

CORPORATE ENGINEERING SERVICES ROLE

Corporate Engineering Services (CES) will provide advice and counsel as appropriate on the health and safety of Company products and services.

MANDATORY MANAGEMENT ACTION

In addition to the broad Company Policy requirements, a minimum number of mandatory actions, as noted below, are considered necessary to assure and maintain an effective Company response to health, safety, and environmental protection obligations:

- Serious health incidents must be reported promptly to the Company Medical Director.
- Component management shall promptly forward to the Corporate Managers of Employee Health, Safety, and Environmental Protection information on
  —work-related employee fatalities and other serious safety incidents requiring a report to a government agency
  —injury and illness experience data, drawn from company logs and records
  —brief summaries on appropriate forms of agency inspection and citation experience, investigations, and governmental proceedings alleging violations of health and safety statutes or regulations
  —brief summaries of agency inspections, serious environmental incidents requiring a report to a government agency, and governmental proceedings alleging violations of environmental statutes or regulations
  —substantial risks connected with the Component's participation in hazardous business activities
- Component counsel shall promptly forward to the Corporate Counsel brief summaries of legal proceedings alleging significant property damage or personal injury from environmental contamination or exposure to hazardous substances.
- Components shall conduct formal health and safety reviews for new or redesigned products, considering the intended application and

PROCEDURE FOR IMPLEMENTATION OF HEALTH, SAFETY, AND ENVIRONMENT POLICY	**PAGE:** 3 of 3

reasonably forseeable circumstances of use, including the effect of customer instructions and warnings.

- Components will conduct systematic monitoring of product safety performance to determine whether hazards exist, and whether they are being eliminated or minimized as intended by Company Policy. Components will also report significant product or service health and safety developments to the responsible operating officer.

REPORTS BY CORPORATE HEALTH, SAFETY, AND ENVIRONMENT OPERATIONS

Regular experience reporting by CHSEO will include:

- illness and injury and workers' compensation experiences for domestic and international operations
- a compilation of Company government agency experience to components
- communication of work-related fatalities to the Corporate Executive Office
- a compilation of environmental protection inspection experiences to components

REPORTS BY CORPORATE MEDICAL

Regular experience reporting by Corporate Medical will include:

- summary of Corporate Medical annual business reviews
- reports on employee and occupational health matters, to be prepared as required under the direction of the Company Medical Director

REPORTS BY CORPORATE ENGINEERING SERVICES

Reports on the safety of Company products and services will be prepared as required under the direction of the Vice President of Corporate Engineering Services.

RESPONSIBILITY FOR COUNSELING

- Vice President, Corporate Health, Safety, and Environment Operations
- Company Medical Director, Corporate Medical
- Corporate Manager, Health, Safety, and Environmental Protection
- Manager, Occupational Hygiene
- Manager, Occupational Safety
- Manager, Environmental Protection
- Manager, Hazardous Materials Transportation
- Vice President, Corporate Engineering Services

2

Program Guidelines and Recommended Practices

INTRODUCTION

An effective occupational hygiene program begins with the recognition of potential physical (noise, heat, and radiation) and chemical health hazards in the workplace. (See Appendix B.) This includes acquiring volume, usage, and hazard information on plant chemicals. It also includes reviewing plant layout, understanding each process, and developing exposure-related medical surveillance information. These steps help guide the occupational hygiene program in the most effective and cost-efficient manner.

After the extent of potential hazards is determined, the degree of hazard is often evaluated quantitatively, using personnel exposure monitoring equipment, sound level meters, light meters, and the like. Reliable exposure monitoring requires understanding and use of correct sampling methods, attention to detail, and an understanding of work processes. The exposure evaluation may suggest that a problem does not exist or may indicate the need for controls to reduce exposures to acceptable levels.

Available exposure control methods include material substitutions, process or work process changes, engineering controls, and use of personal protective equipment. Exhaust ventilation is a common type of engineering control for chemical exposure. Noise damping, isolation, and process redesign are also examples. Personal protective equipment is the least acceptable alternative, because the source of the exposure problem remains. Such equipment must also be carefully evaluated to ensure that its use provides adequate exposure protection. Regardless of its type, employees need training in the use and care of personal protective equipment.

Employees must be informed of the potential hazards of the workplace. Necessary information includes the nature of chemical and physical hazards, sampling results, means to prevent exposures, and emergency response plans. A company-wide hazard communication program can assist with this effort. (Hazard communication guidelines are provided later in this section.) Finally, all aspects of the occupational hygiene program should be clearly documented so that this information can be used to its fullest advantage.

ADMINISTRATION

Implementation

Guideline: The fundamentals of the occupational hygiene program outlined in this manual should be implemented at all places of business, including manufacturing, service, construction, and laboratory locations.

Recommended Practices

- The site manager should be responsible for implementing and maintaining the occupational hygiene program.
- For effectiveness, a particular site may require the services of an occupational hygienist.
- If the hygienist is a company employee, he or she should be part of the management team and involved in pertinent business decisions. He/she should be provided with continuing education opportunities to keep abreast of current science. If consulting occupational hygienists are used, they should be highly competent and, where possible, professionally licensed and/or board-certified.

- All levels of management should understand and accept their responsibilities for minimizing health risks.

Responsibilities

Guideline: All salaried and hourly employees have responsibilities for health and safety.

Recommended Practices

- Clearly define specific responsibilities. Use position guides and other methods to communicate responsibilities within the organization as appropriate. While specially trained professionals (occupational hygienists) should be responsible for technical aspects of the program, not all activities require occupational hygiene professionals. Managers at all levels should be responsible for implementing occupational hygiene policies and programs to an extent consistent with their positions within the organization.
- Contact company professionals within the business in the specific location's executive office or corporate office as needed for advice, counsel, and assistance in the interpretation of local regulations.
- Develop procedures to ensure that proper safe practices are followed by all personnel and contractors when working at job sites off company premises.
- Develop procedures to ensure that onsite contractors working within the confines of any facility adhere to company health and safety requirements.
- Develop procedures that outline actions to be taken in specified emergency situations to protect the health of employees and the local community.

Documentation

Guideline: All relevant legislation, guidelines, documented procedures, and practices should be kept onsite and made available to employees as required.

Recommended Practices

- Obtain copies of all relevant regulations and guidelines. These materials should be posted and/or made available to employees as required by applicable laws and regulations.
- Develop a detailed occupational hygiene program. This manual may be used to prepare such a program.
- Prepare written procedures for all potentially hazardous situations. (See Appendix A.)
- Prepare a site emergency plan. All employees should be trained in, familiar with, and/or have access to the plan.
- As required by law, make documentation available to government agency personnel upon request.

  Note: Company administrative procedures should be followed when handling requests. Completion of appropriate forms and logs may be required by the corporate health and safety office.
- Prepare all occupational hygiene documentation and reports in a professional manner, avoiding speculation and including only factual information. Make appropriate information available to employees and government agency personnel as required by company bulletins, directives, and applicable regulations. Maintain occupational illness records in the facility medical files and related personnel folders.
- Record work-related injuries and illnesses on approved government logs and forms (e.g., in the United States, OSHA Form 101 and Log 200).

RECOGNITION

Guideline: Exposure to chemical and physical agents can result in harm to employees. Recognition of situations and conditions that may present exposure risks (based on usage, handling, controls, and the nature of the materials or energy sources) is the first element of an effective occupational hygiene program. Recognition can be achieved through observation of usage and handling and a knowledge of the chemical materials or energy sources involved.

Recommended Practices

- Prepare a plant layout to show process equipment including hazard control devices, material storage and handling areas, and eating and hygiene facilities. The plans should include all storage tanks (underground or surface), related piping systems, and energy sources. Diagrams should be updated as processes change. Maintain all plans as part of the permanent occupational hygiene record.
- Develop a list of chemicals and energy sources used or produced for each work area. The inventory should include locations, chemical volumes used per year, maximum quantity on hand, Material Safety Data Sheet (MSDS) or similar document reference numbers, and location and description of energy sources. This inventory should be updated at least annually; it is integral to the development of an effective occupational hygiene program.
- Acquire MSDSs or similar documents for each chemical in the inventory. Equipment specifications should be available for all energy sources.
- Develop quality MSDSs and equipment specifications for company products. (Consult the company hazard communication plan for further information.)
- Conduct a Job Hazard Analysis (JHA) for each job, as appropriate.
- Undertake a walk-through survey. Note all relevant information that may impact on health (e.g., processes, products, by-products, energies, use and quality of control systems, and employee work procedures). Have a professional occupational hygienist conduct this survey and/or review all collected information.

- Appropriately label all containers, pipes, conduits, storage vessels, etc.
- Research and/or develop medical information and records on all employees to identify any high-risk individuals or groups. Ensure that medical facilities are aware of all potential risks.

EXPOSURE ASSESSMENTS

Guideline: Subsequent to completing a comprehensive hazard recognition review, each facility should conduct an exposure assessment. An occupational hygienist should be responsible for developing and implementing these evaluations.

Recommended Practices: General

- Develop a list of materials and energy sources that must be evaluated. Identify priorities. Advise management on how the evaluation will be conducted.
- Exposure assessments, air monitoring, and surveys should be conducted in accordance with standard occupational hygiene practices. Samples should be collected and analyzed using approved methods (e.g., National Institute of Occupational Safety and Health Physical and Chemical Analytical Methods [NIOSH P&CAM]).

Note: If an evaluation is not conducted because of low toxicity or low potential for exposure, document the reasons for those conclusions.

Recommended Practices: Sampling

- Sample for a minimum of four hours or for a period required by a recognized methodology or regulatory requirement.
- Sample during a period that is representative of typical exposures. Also sample to characterize exposures during unique operations and/or conditions.
- Sample initially as frequently as is needed to establish

time-weighted average (TWA) exposures, and then sample periodically to ensure continuing control.

- Use a statistical method to ensure that samples are representative and will meet regulatory compliance or scientific requirements.
- Use personal sampling in preference to area sampling for employee exposure evaluations.
- Sample individuals with potential for greatest exposure. Sample other employees doing similar work to obtain representative data. Use all data for compliance evaluation.
- Where appropriate, use direct reading instruments that are approved and calibrated against appropriate standards.

  Note: Direct reading area monitors may not represent true personal exposure.
- Maintain all equipment, systems, and monitors according to manufacturer's recommendations.
- Calibrate all equipment, monitors, and samplers that can be field-calibrated before and immediately after a testing program. Quantitative laboratory calibration must be undertaken according to standard practices and/or equipment specifications. Record all quality assurance and maintenance data.
- Use approved sampling and analytical methods (e.g., NIOSH) where appropriate. If possible, use only company-approved laboratories. (See "Use of Support Laboratories/Consultants," later in this section.)
- Where appropriate, prepare sampling blanks and handle accordingly. Occasionally use sample splits to verify laboratory accuracy and reliability.
- Label and store samples properly to avoid loss or damage.
- Record data in accordance with guidelines discussed under "Recordkeeping," later in this chapter.

- Retain records in accordance with company and regulatory requirements.
- Provide exposure data to employees and to the company medical center. As required by applicable regulations, make data available to regulatory personnel upon request.
- Initiate additional control procedures or practices where data shows exposure approaches or exceeds permissible limits.

  Note: A level of 50% of a permissible limit is generally recommended as an action level to implement controls.

Recommended Practices: Physical Hazard Evaluations

- Develop an inventory of physical hazards (e.g., noise sources or radiation sources such as X-rays, radioactive materials, or lasers).
- Apply a statistically valid methodology to obtain representative and/or worst-case exposure measurements.
- Ensure proper calibration per manufacturer's recommendations and/or regulatory requirements.
- Use equipment in conformance with manufacturer's instructions.
- Record and analyze data systematically. (See ''Recordkeeping,'' later in this section.)
- Retain records in accordance with company and regulatory requirements.
- Have all exposure assessment data and information reviewed, analyzed, and interpreted by an occupational hygienist.

Note: Regulations or company programs may designate some chemical or physical agents for special assessments because of their potential to cause significant harm. (Check relevant company bulletins and guidelines.)

HAZARD REDUCTION PROCEDURES

General

Guideline: Hazard reduction can be achieved by a variety of individual methods or by a combination of methods. All materials and energy sources must be handled and used in such a way that the health of workers and the community is not jeopardized. Every attempt should be made to reduce exposures to the lowest practical limits.

Note: Personal protective devices are a *last line of defense* in exposure control. Appropriate engineering, administrative and work practice controls, and product or process substitutions should be tried before personal protective devices are used.

Recommended Practices

- When selecting control methods, consider the specific hazard, work, processes and number of employees involved, the work environment, available control alternatives, and all exposure and sampling data.
- Consider all possible routes of exposure when developing and implementing prevention programs.
- Eliminate hazards where possible. In some cases, substitution of a less hazardous process or material will considerably reduce the potential for harm. Ensure that any substituted environment, method, or chemical is safer than that originally used.
- When possible, reduce hazards by process redesign, preventive maintenance, or equipment modification.
- When a hazard cannot be eliminated, control at the source of generation is usually the most effective means of exposure reduction. Evaluate appropriate engineering controls, such as local exhaust ventilation, to reduce airborne hazard potentials at or near the source of generation.

- Energy sources, especially noise, radiation, lighting, heat, and vibration, may need special control procedures. (See "Physical Hazard Controls," later in this section.)
- Examine and modify employee work practices as necessary to decrease exposure and hazard potentials.
- Use personal protective devices as a last resort or as an additional control measure, if the foregoing methods prove to be unsuccessful, inappropriate, infeasible, or inadequate.
- Train employees to understand hazard potentials and the importance of appropriate control methods.

Note: Material Safety Data Sheets provide information for minimizing exposures. Procedures should be followed as appropriate. Production personnel should provide supplementary information detailing proper use and control of materials in their particular operation.

Ventilation Control

Guideline: Good ventilation practices must be applied when designing, operating, and maintaining systems. The "Ventilation Manual of Recommended Practices" prepared by the American Conference of Governmental Industrial Hygienists (ACGIH) is a recognized source document for ventilation control procedures and practices.

Recommended Practices

- Document and maintain design and performance specifications for the ventilation system. (Research of existing facility engineering files can assist this effort.)
- When local exhaust systems are used for materials that are considered toxic or very toxic, ensure the following:
  1. The capture velocity is adequate to capture and transport the contaminant.
  2. The air flow pattern draws the contaminant away from the employee's breathing zone.

3. The equipment material and construction are compatible with the contaminant.
4. The fan size and rating are sufficient to meet system operating needs.
5. Appropriate air cleaning systems are installed to prevent air pollution.
6. Adequate, tempered makeup air is provided.
7. Roof exhaust and intakes are located so as to avoid reentrainment of contaminated air.

- Develop and implement an operation and preventive maintenance plan for all ventilation systems, and maintain appropriate documentation.
- All critical flow systems for highly toxic materials must be monitored against design specifications at least monthly; monitoring data must be recorded. Other systems should be monitored as frequently as is necessary to ensure continued integrity; monitoring results should be recorded.
- Air recirculation systems must be checked frequently to ensure clean air delivery.

Note: In some cases, general exhaust or dilution ventilation can be used for controlling materials of low toxicity. The location of the supply and exhaust diffusers should be designed to minimize potential employee exposure and to provide appropriate air flow patterns. Professional guidance should always be sought when considering this alternative.

Personal Protective Equipment

Guideline: It is prudent for each location to develop and implement written procedures to address the purchase, approval, use, and maintenance of protective equipment. More detailed procedures for a respiratory protection program may be required.

Recommended Practices

- Identify areas, operations, and environments where protective equipment and/or devices are required.
- Develop written procedures regarding the selection, use, procurement, inspection, and maintenance of protective equipment used by employees and visitors. Management should ensure that the required equipment is available and is distributed as appropriate to all visitors entering the plant area.
- Develop and implement procedures to ensure that protective equipment is purchased only with the approval of the local health and/or safety authority. No substitutions should be permitted without the prior written approval of the local health and/or safety authority.
- The local health and/or safety authority must ensure that all emergency, safety, or protective equipment is regularly inspected, properly maintained, and ready for immediate use.
- Ensure that employees are instructed and trained in the proper use, limitations, and care of the equipment and devices they are required to use; this is usually the responsibility of a supervisor. Maintain complete records on this training.
- All protective equipment, as required by location procedures, should be readily available. This may include the following:

    1. eye protection
    2. face protection
    3. head protection
    4. hearing protection
    5. protective clothing
    6. respiratory protection

7. safety belt/harness
8. safety shoes
9. skin protection (gloves, apparel, creams, lotions, etc.)
10. other equipment required for specific operations

- Ensure that all protective equipment and/or devices meet applicable government regulations or recognized consensus standards.

Note: Medical considerations may impact on an employee's ability to wear personal protective equipment. *All* employees who use respirators should be evaluated annually for capability to work with a respirator. A statement of fitness completed by a physician should be available in the employee's file.

Physical Hazard Controls

Guideline: Each location is required to develop and implement written procedures outlining the precautions necessary for the protection of all workers exposed to physical hazards (noise, vibration, light, temperature extremes, and radiation).

Recommended Practices

Noise

- Identify and document significant noise sources and noisy work areas with sound pressure level (SPL) greater than 80 decibels, A-weighted (dBA).
- Consider all potential sources of noise (e.g., fan blades, air ducts, valves, and other mechanical equipment) and consider alteration to reduce noise levels (e.g., isolation, enclosure, dampening, or insulation)
- Alter or enclose noise sources where appropriate to reduce noise at the source.
- Where appropriate, use absorption walls or treat walls, floors, and ceiling with noise-absorbing materials.

- Isolate noise sources from work areas (e.g., use isolation mounts to reduce transmitted noise to floor and equipment).
- When there is a demonstrated need, use approved hearing protection.

Note: In some cases it is advisable to seek the services of an occupational hygienist or professional experienced in acoustics to evaluate the frequency spectrum of the noise source and recommend controls. It is also advisable to provide employees working in high-noise areas (>80 dBA) with a professional audiometric evaluation annually to detect temporary or permanent threshold shifts from the last recorded hearing baseline.

Vibration

- Identify and document all equipment and/or processes that present a vibration hazard (e.g., operation of vibrating tools or power equipment).
- Consider substitution or modification of equipment or processes to reduce vibration hazards.
- Use isolation where appropriate to reduce transmitted vibration (e.g., using flexible connections).
- Change energy and/or frequency of vibration where appropriate. Low frequencies (25–75 hertz) are more damaging than higher frequencies. (See Appendix B.)

Lighting

- Identify and document all processes or tasks that have inadequate lighting or special lighting requirements.
- Consult applicable regulations or guidelines for lighting evaluation (e.g., American National Standards Institute [ANSI] Practice for Industrial Lighting).
- Make appropriate measurements of light intensity, luminance, and reflectance. Consider the type of work area as well as tasks performed.

- Minimize glare and surface reflection by proper design, surface preparation, and selection of non-glare materials.
- Maintain all fixtures to original design specifications.
- Provide localized lighting for fine tasks where appropriate. Visibility of the task or object should be evaluated by size, contrast, time of viewing, and brightness.
- Use colors to provide a comfortable glare-free environment with appropriate contrast.
- Where appropriate, shade natural light to minimize glare.

Heat Stress

- Identify and document all areas, environments, or jobs with potential heat exposure hazards.
- Evaluate all components that contribute to heat stress (e.g., radiant or infrared heat, humidity levels, metabolic load or work rate, air velocity, and air temperature).
- Make appropriate heat stress measurements (e.g., Wet Bulb Globe Temperature—see Appendix F).
- Use appropriate techniques to minimize heat stress. (For example, insulate sources, decrease or increase ventilation rates, or use diffusers, dehumidifiers, or reflective surfaces).
- Use administrative controls, such as rest breaks, as appropriate.
- Provide rest areas away from extremely hot environments.
- Provide easy access to cool drinking fluids.
- Provide medical coverage/attention as appropriate.
- Provide training to employees. Where appropriate, institute acclimatization schedules.
- Provide appropriate clothing as warranted by heat source and heat stress components.

Cold Stress

- Identify and document all areas, environments, or jobs with potential cold temperature exposure hazards.
- Evaluate all components that contribute to cold stress (radiant loss to cold surfaces, metabolic load or work rate, air velocity, or air temperature).
- Make appropriate measurements (e.g., air temperature, velocity, and wind chill).
- Use administrative controls, such as rest breaks, as appropriate.
- Provide rest areas away from extremely cold environments.
- Provide easy access to hot drinking fluids.
- Provide medical coverage/attention as appropriate.
- Provide training to employees.
- Provide appropriate clothing as warranted by exposure.

Radiation (Ionizing and Non-Ionizing)

- Develop procedures that require specific programs for the protection of workers at individual locations who may be exposed to radiation. These procedures should address the level and type of radiation and meet all company standards and government regulations for protection. Procedures should include the following:

    1. Identify and train as necessary a radiation protection officer if appropriate or required by local or national regulations.
    2. Identify and document all radiation sources and materials.
    3. Use radiation dosimeters (where applicable) or other radiation monitors to evaluate employee exposures and maintain adequate records of exposure.

4. Minimize the number of individuals exposed by restricting access to radiation areas and radioactive materials or sources.
5. Use radiation work permits (RWP) where suitable.
6. Reduce exposure where possible by isolating or enclosing equipment or materials, providing remote operation, and reducing time of exposure.
7. Ensure sources are designed and equipped with adequate protective devices (e.g., shields, interlocks, warnings, and automatic shutdown mechanisms) and that these devices are periodically evaluated and maintained according to manufacturer's instructions.
8. Provide appropriate labeling and warnings on all equipment and sources.
9. Develop and institute good work practices that minimize exposure and potential contamination.
10. Develop emergency procedures to minimize exposure, release, and contamination.
11. Ensure compliance with regulations or guidelines pertaining to environmental protection and waste disposal.
12. Utilize protective clothing and equipment to reduce exposure.

Note: Most countries have specific regulations or guidelines for the handling, use, storage and disposal or radioactive sources. You should be familiar with local laws.

HAZARD COMMUNICATION

Guideline: U.S. operations must comply with the OSHA Hazard Communication Standard (29 CFR 1910.1200). Operations outside the United States should develop and implement a similar program or a program that meets appropriate regulatory requirements. The Hazard Communication Standard requires employers to evaluate the hazards of materials produced or imported and to communicate such hazards to all users and employees.

Recommended Practices

- Adopt or develop a written hazard communication plan.
- Assign a person or persons to be responsible for ensuring compliance with the plan, including inventory development and updating, proper labeling of containers, procurement and management of MSDSs, and training of employees.
- Prepare a list of materials manufactured or imported and define associated hazards. Note and archive source documents used in preparing the data.
- Develop an MSDS and/or equipment specifications for appropriate products.
- Develop and implement an employee training program to ensure that all employees are fully aware of the chemical and physical hazards associated with the working environment.

TRAINING

Guideline: It is management's responsibility to ensure that *all* employees, including new hires, are trained and are able to perform their work in a healthy and safe manner.

Recommended Practices

- Designate responsibilities for developing and conducting health and safety training.
- Identify training needs by considering work operations or processes, general work environments, materials handled or used, potential exposures to all chemical or physical agents, emergency response procedures associated with the work operations, and other factors as appropriate.

- Develop written procedures to ensure compliance with training needs.
- Regularly reassess training needs and prepare a specific plan to meet ongoing training requirements.
- Maintain a complete record for each employee receiving health and safety training.
- Implement specific programs for the training of personnel involved with regulated or hazardous work activities or environments.
- Ensure that each location has personnel trained in first aid and cardiopulmonary resuscitation (CPR).
- Train all employees in fire safety and emergency programs.
- Retrain employees as required (i.e., transferees, employees returning to work after an extended absence) to maintain proficiency and knowledge.

RECORDKEEPING

Exposure Records

Guideline: It is necessary to develop exposure data records for chemical and physical agents to assist in the process of recognizing or identifying disease potentials and satisfying regulatory requirements.

Recommended Practices

- Develop occupational hygiene records to comply with accepted standards of practice regarding content of the records, data to be included, and length of retention.
- File, maintain, and make available for inspection, as appropriate, all exposure records. Copies should be available to the responsible medical operation and maintained with appropriate employee medical records.
- Post all summary results of exposure monitoring or inform those workers directly affected by the testing, as

required by applicable regulations. (Some national regulations require employee notification within a specific time period.)

- Maintain records for periods specified by company procedures and/or as defined in applicable regulations.

Notes: (1) Occupational hygiene data may be needed in legal situations (e.g., toxic tort cases); consequently, records should be clear, accurate, well-documented, dated, and signed. (2) Maintenance of records should be the responsibility of the appropriate local or professional management; medical records should be under the control of a medical professional and kept confidential.

Accident/Illness Reporting and Recordkeeping

Guideline: All locations should record *all* occupational illnesses, injuries, and accidents and report serious incidents to local agencies as required by law. Statistics should be maintained and analyzed periodically to identify problem areas requiring management attention.

Recommended Practices

- Ensure that all logs and records required by regulations are properly maintained and available for inspection at each business location.
- Comply with all applicable laws, regulations, and company procedures for recording and reporting all occupational injuries and illnesses at each location.

USE OF SUPPORT LABORATORIES/CONSULTANTS

Guideline: Occupational hygiene consultants and laboratories used for analyzing samples to meet the needs of this hygiene program must be of the highest quality.

Recommended Practices

Laboratories

- Ensure that all laboratories used are certified by a recognized health/hygiene authority. (When using laboratories accredited by the American Industrial Hygiene Association [AIHA], make sure that they participate in proficiency testing for the analyses required.)
- Use split samples to check the validity and accuracy of data.
- Investigate laboratory practices whenever findings appear to be inconsistent. For example, visit the site to ensure proper handling, analysis, quality assurance, and recording of samples.
- Handle all samples carefully to prevent leakage, cross-contamination, breakage, loss, or misidentification.
- Properly label all samples.

Consultants

- All occupational hygiene consultants should be certified by a recognized authority and/or have the necessary qualifications, training, and experience to handle a given project.
- The consultant must follow all relevant practices in this program.
- Seek references from previous users of the consultant as to the quality of work performed.
- Sign confidentiality agreements with consultants forbidding the disclosure of company data and information to any third party without written consent of management.

3

Auditing and Measurement

INTRODUCTION

Program evaluation is an integral part of an effective occupational hygiene program. Evaluations can and should be performed by the occupational hygiene professional or program manager on a regular basis. The evaluation process should consider not only the physical aspects of facility hazards and control procedures but also administrative procedures, including training and documentation.

AUDITING AND MEASUREMENT PROCEDURES

Guideline: To measure the effectiveness of an occupational health program, self-evaluation or self-appraisal is critical. Four basic areas must be evaluated against the program's goals.

Recommended Practices

- Frequently review the program's organization and procedures to ensure adequacy and efficient use of resources.
- Evaluate occupational illness experience to identify trends and to ensure that controls are effective.
- Hazard identification, evaluation, and control programs should be reviewed frequently for technical competence.

- Evaluate training and communications procedures to ensure they address the needs of current, new, and transferred employees; modify as necessary.

ORGANIZATION AND PROCEDURES

Guideline: The people that comprise a successful health and safety program should be evaluated for their ability to prevent incidents, attain stated goals, and adapt to changes in program focus.

Recommended Practices

- Screen health and safety programs for content and applicability.
- Review emergency procedures and the adequacy of communications to affected employees and other personnel.
- Review illness recordkeeping practices to ensure compliance with company procedures and regulatory requirements.
- Document the program's organization and procedures. Make sure that the program is maintained between review periods. To be effective, the program should be actively supported by the *highest management level* in the organization and must gain attention, understanding, and acceptance from the employees for whom it is written.

OCCUPATIONAL ILLNESS AND INJURY EXPERIENCE

Guideline: The evaluation of all occupational illnesses and injuries is an essential measurement of the effectiveness of employee health programs.

Recommended Practices

- Review and analyze data to identify trends and patterns. Use this analysis to detect unsafe conditions or workplace practices.
- Analyze the costs associated with occupational illnesses. This will highlight the savings that can be gained through the prevention of employee exposure to chemicals and physical agents.

Note: Workers' compensation programs can vary significantly from country to country and even within countries. A facility's loss history can be used to measure a program's effectiveness, against which future performance can be compared.

HAZARD IDENTIFICATION AND CONTROL

Guideline: Facilities should regularly review the hazard identification, evaluation, and control components of the employee health program.

Recommended Practices

- Review exposure survey results for quality, completeness, and trends.
- Review hazard recognition procedures and documentation (e.g., inventories, walk-through surveys, process reviews, area plans, and MSDS files).
- Review job hazard analysis procedures, including who conducts the analysis and the frequency of analysis.
- Review employee feedback systems, including style, usage, value, and management follow-up.
- Review adequacy of process and equipment design to control or minimize exposures.

- Ensure that standard operating procedures include health control information. These procedures should be reviewed by competent professionals prior to their release to manufacturing.
- Review preventive maintenance programs for control systems and equipment.
- Review personal protective equipment procedures, assignment, maintenance, control, and use.
- Review management audits, including personnel involved, type, frequency, and follow-up procedures or systems.
- Review employee training programs, including type, frequency, professionalism of trainers, and completeness. (See "Training and Communications," below.)
- Review emergency response capabilities of in-plant personnel and equipment; examine relationships with local communities and agencies.

TRAINING AND COMMUNICATIONS

Guideline: Employee participation is integral to an effective training program. These programs should be reviewed for relevance to particular job assignments; success in informing, motivating, and modifying employee behaviors; and ability to convey management's commitment to a healthy work environment.

Recommended Practices

- Review procedures for informing new employees on the health program's procedures and application.
- Review hazard communication training, both in the classroom and at the work station. Confirm employee understanding.

- Review job training programs, both formal and on-the-job (OTJ) training.
- Review written communications, including style, type, and frequency.
- Review frequency and content of employee meetings and awareness programs.
- Review mechanisms to facilitate employee feedback, such as suggestion systems, round-table meetings, and management lunches.
- Review professional training skills of the "trainers" (employee relations staff, production management, engineers, and supervisors).

Appendix A

Typical Program Elements

GUIDELINES

Items and issues to be covered in a comprehensive occupational hygiene program will vary with the type of facility and operation. Applicable guidelines and practices need to be developed for each major and specific occupational hygiene element.

RECOMMENDED ELEMENTS

- Airborne contaminants: exposure assessment and control
- Asbestos: assessment and control
- Building indoor air quality
- Carcinogens, teratogens, mutagens, and reproductive health
- Cold stress: exposure assessment and control
- Compressed gases: storage and use
- Confined space entry
- Control strategies: administrative and engineering controls
- Decontamination procedures
- Employee training and hazard communication

- Ergonomics (work station design)
- Flammable and combustible liquids: storage and use
- Hazardous materials: storage and use
- Health and safety agency liaisons and inspections
- Heat stress: exposure assessment and control
- Human factors engineering (equipment design)
- Job hazard analysis (JHA)
- Laboratory procedures and sample control systems
- Maintenance of controls and facilities
- Noise exposure: assessment and control
- Occupational exposure limits and guidelines
- PCBs: assessment and control
- Personal protective equipment
- Preventive maintenance programs
- Radiation: ionizing and non-ionizing
- Recordkeeping and reporting
- Respiratory protection
- Sampling strategies: personal and area monitoring, contaminated surfaces
- Self-analysis, measurement, and appraisal
- Ventilation controls
- Video display terminals: ergonomic aspects

Appendix B

Chemical and Physical Agents: Descriptions and Sources

This appendix provides a brief overview of common chemical and physical agents present in the work environment and gives an indication of the adverse effects possible from exposure. The list given here is not exhaustive. Specific hazards present are identified based on recommendations presented in Section 2.

1. **Chemical Agents:** Chemical agents include elements, compounds, or mixtures. They may occur in airborne, liquid, or solid form. Exposure may occur through inhalation, absorption, or ingestion. However, inhalation of airborne forms (e.g., dusts, fibers, fumes, smoke, vapors, and gases) is the most common route of exposure in the industrial environment. Exposure may result in a wide variety of adverse health effects. (See Appendix C.) Effects may be localized or systemic, acute or chronic.

   Sources: There are numerous sources of these agents throughout the work environment, including raw materials or products used, finishing products, process by-products, intermediates or other emissions, building products, cleaning products, and the like. When evaluating a chemical agent, the primary ingredient, as well as additives and/or impurities, should be considered; the manner of use and time of use are also integral factors in exposure assessment.

2. **Noise:** Excess noise may be considered as sound pressure levels consistently above 80 dBA. Most current regulations limit noise exposure when it reaches a level equivalent to 85 dBA for an eight-hour period. Excessive exposure to noise may result in temporary or permanent hearing loss. Damage usually occurs first at high frequencies, reducing the ability to hear noise in the frequency region.

   Sources: Again, numerous sources of noise exposure may be present in the industrial environment, including power tools, vibrating equipment, pneumatic tools, blowers, fans and duct work, and compressors. When evaluating noise exposure, it is important to consider the intensity (sound pressure level) and frequency of the noise as well as the duration of exposure.

3. **Vibration:** Physical hazards due to vibration are a result of the rapid motion of vibrating equipment transmitted to the body. Generally, the hands are affected. A condition known as Raynaud's Syndrome ("white finger") may develop, mainly in the fingers. This effect is a result of impaired circulation.

   Sources: This condition is generally limited to occupations using power vibrating tools such as air hammers, chisels, electric saws, pneumatic tools, and the like. When evaluating exposure, frequency of vibration as well as duration of exposure are important factors. Low frequencies (25 to 75 hertz) are more damaging than higher frequencies.

4. **Non-Ionizing Radiation:** Non-ionizing radiation does not have sufficient power to produce ionization in interaction with matter. The effects are primarily thermal. However, photochemical effects may occur at some frequency ranges. Depending on the wavelength and source of the energy, the eyes and/or skin will be the primary organs affected. Effects are listed in Appendix C.

Sources: Some common sources of non-ionizing radiation are provided below:

- ultraviolet radiation—natural light, artificial lighting, fluorescent lights, electric welding arcs, lasers
- visible light—high-intensity light sources, lasers
- infrared radiation—high-surface temperature equipment or material (usually metallic); driers, finishers, heaters designed to operate at these frequencies; lasers
- microwaves and radio waves—FM, TV, radar transmitters and antennas; medical devices; microwave ovens; communication towers; radar equipment
- radio frequency—radio or high-frequency electrical heating equipment, induction and dielectric heaters

5. **Ionizing Radiation:** Ionizing radiation is capable of ionization (production of ion pairs) in interaction with matter and can cause irradiation, alteration, or destruction of body cells, depending on the intensity, type of radiation, and time of exposure. Acute and/or chronic effects may occur from overexposure. Effects may be direct to the exposed individual or genetic. The major types of ionizing radiation include alpha, beta, gamma, X-ray, and neutron. Hazards may be presented by external and/or internal exposure when radioactive equipment or materials are utilized. Detailed and comprehensive health and safety procedures must be developed and implemented.

   Sources: Cosmic radiation, nuclear materials, radioactive source materials, high energy accelerators, medical equipment, analytical laboratory equipment.

Appendix C

Toxicology

INTRODUCTION

This section summarizes the principles of toxicology and its relationship to the field of occupational hygiene.

Toxicology is the study of the adverse effects of chemicals on living organisms. Toxicologists are trained to determine the nature of these adverse effects and to assess the probability that a specific exposure will cause a harmful effect. Occupational hygiene concerns itself with identifying and minimizing exposures in the work environment that may cause an adverse effect. Toxicologists and occupational hygienists work cooperatively to prevent adverse health effects from occurring in workers exposed to workplace chemicals.

BACKGROUND

Toxicology is a very broad science. It deals with chemicals used in industry, agriculture, drugs, food, and cosmetics, as well as those occurring naturally in the environment. The primary objective of industrial toxicology is the prevention of adverse health effects in workers exposed to chemicals in the workplace. It is the toxicologist's job to determine how and where in the body a substance will exert its toxicity and to determine, with some degree of confidence, the level of exposure at which the risk of adverse health effects is negligible. The occupational hygienist uses the toxicity information to prescribe safety measures for protecting workers.

IMPORTANT PRINCIPLES OF TOXICOLOGY

1. Toxicity vs Hazard

It is important to differentiate between *toxicity* and *hazard* or *risk*. *Toxicity* is the intrinsic capacity of a chemical to cause harmful effects to humans and other living organisms. It is an inherent property of a material, like flash point. In practical situations, the hazard or risk associated with the use of a substance is more important than the toxicity. *Risk* is the probability that a substance will produce harm under specific conditions. *Safety*, the opposite of risk, is the probability that harm will not occur under specific conditions. The application of certain procedures that minimize the hazard or risk of using a chemical ensures safe working conditions. A very toxic chemical may be nonhazardous when used under safe conditions.

2. Dose-Response Relationship

In the strict use of the term, the dose of a chemical to the body resulting from exposure is directly related to the degree of harm. Assuming this straightforward relationship allows the toxicologist to determine a threshold level of exposure for a given chemical. The threshold level is the highest amount of a chemical to which one can be exposed with no resulting adverse health effect.

3. Routes of Exposure

For a chemical substance to produce a harmful effect, it must reach the appropriate site in the body at a concentration and for a length of time sufficient to produce an adverse effect. There are three major routes by which chemicals enter the body in the workplace.

Inhalation is the most important route of entry for harmful substances in industrial exposures. Dusts, fumes, mists, gases, vapors, and other airborne substances may enter the body via the lungs and may produce local effects on the lung or may be transported by the blood to specific organs in the body.

Contact is another important route of exposure for harmful substances in industrial exposures. The skin may act as a barrier to some harmful agents, but other materials may irritate or sensitize the skin and eyes or travel through the skin into the bloodstream, thereby impacting on specific organs.

Ingestion may occur directly or indirectly. The material enters the intestines and is absorbed into the bloodstream. Industrial exposure to harmful substances through ingestion may occur when workers eat lunch, drink coffee, apply cosmetics, or smoke in a contaminated work area. The materials may exert their toxic effect on the intestinal tract or at specific organ sites.

4. Target Organs

Chemicals and other substances may exert their harmful effects on specific organs of the body, such as the lungs, liver, kidneys, central nervous system, and skin. These specific organs are termed target organs and will vary with the chemical of concern.

5. Short-Term vs Long-Term Effects

Short-term (*acute*) adverse effects may occur within a short time period following either a single exposure or multiple exposures to a chemical.

Long-term (*chronic*) adverse effects usually occur after a long period of time following exposures to small quantities of a chemical. Chronic effects may sometimes occur following short-term exposures.

6. Effects from Selected Occupational Agents

Table C.1 shows the harmful effects that can result from overexposure to some chemical agents. Table C.2 shows the harmful effects from some physical agents.

Table C.1 Harmful Effects Resulting from Overexposure to Selected Chemical Agents

Type of Stress or Agent	Major Route of Exposure	Potential Effects of Overexposure (Acute/Chronic)
Dusts		
Asbestos	Inhalation	*Chronic:* asbestosis, lung cancer, mesothelioma (cancer of the squamous epithelium)
Crystalline silica (quartz)	Inhalation	*Chronic:* fibrosis of lung
Metal Dusts and Fumes		
Arsenic	Skin contact	*Acute:* skin irritation, conjunctivitis, sensitization dermatitis *Chronic:* possible epidermal cancer and hyperpigmentation
	Inhalation	*Chronic:* mucous membrane irritation, pulmonary irritation, peripheral nerve damage, digestive symptoms, possible lung cancer
Beryllium	Skin contact	*Acute:* primary or sensitization dermatitis, conjunctivitis
	Inhalation	*Acute:* respiratory irritation, bronchitis, dry cough *Chronic:* shortness of breath, dry cough, nonspecific X-ray features; symptoms may occur after long exposure periods and X-ray indication
Cadmium	Inhalation	*Acute:* (particularly fume) chest pain, shortness of breath, pulmonary edema, digestive effects *Chronic:* kidney damage, emphysema, anemia, mucous membrane irritation
Chromium (as chromic acid and chromates)	Skin contact	*Acute:* contact dermatitis, sensitization dermatitis, chrome ulcers

Table C.1 continued

Type of Stress or Agent	**Major Route of Exposure**	**Potential Effects of Overexposure (Acute/Chronic)**
Chromium (as chromic acid and chromates), continued	*Inhalation*	*Acute:* upper respiratory irritation *Chronic:* lung, pharynx, and chronic nasal irritation; perforated nasal septa; lung cancer (certain insoluble hexavalent compounds); kidney damage
Iron	Inhalation	*Chronic:* benign lung disease (siderosis). X-ray indication not clearly associated with serious symptoms
Lead	Inhalation	*Chronic:* gastrointestinal disturbance, anemia due to red blood cell effects, neuromuscular disorders (wrist drop), kidney disease, and reproductive effects
Mercury	Inhalation and skin contact	*Acute:* lung problems, irritation *Chronic:* nervous system disorder, weakness, diarrhea, memory loss, tremors
Nickel	Skin contact	*Acute:* sensitization dermatitis up to 7 days after initial contact
	Inhalation	*Chronic:* possible risk of cancer of paranasal sinuses and lungs from nickel and certain inorganic compounds
Zinc oxide (fume only)	Inhalation	*Acute:* freshly generated fume causes flu-like symptoms, cough, fast heartbeat
Acid Mists/Gases (e.g., hydrogen chloride, sulfuric acid)	Skin	*Acute:* severe corrosive action
	Inhalation	*Acute:* nose, throat, and respiratory irritation *Chronic:* chronic bronchitis, permanent respiratory tissue damage

Table C.1 continued

Type of Stress or Agent	Major Route of Exposure	Potential Effects of Overexposure (Acute/Chronic)
Vapors		
Chlorinated solvents, (e.g., 1,1,1-trichloroethane, perchloroethylene, trichloroethylene)	Inhalation	*Acute:* central nervous system effects, depression, dizziness, nausea *Chronic:* liver damage
	Skin absorption	*Acute:* defatting/drying of the skin, dermatitis
Aromatic solvents (e.g., toluene, xylenes)	Inhalation	*Acute:* central nervous system effects, depression, narcotic effects *Chronic:* liver, blood system disorders
	Skin absorption	Dermatitis (chronic or acute, depending on individual sensitivity)
Ketones (e.g., MEK, acetone, MIBK)	Inhalation	*Acute:* narcotic-like action; nose, eye irritation; nausea *Chronic:* sensitization
	Skin contact	Dermatitis (chronic or acute, depending on individual sensitivity)
Aldehydes (e.g., formaldehyde, acetyl aldehyde)	Inhalation	*Acute:* eye and mucal irritation *Chronic:* sensitization
	Skin contact	Dermatitis (chronic or acute, depending on individual sensitivity)
Alcohols (e.g., methanol, glycols)	Inhalation	*Chronic:* liver damage
	Skin contact	Dermatitis (chronic or acute, depending on individual sensitivity)
Gases		
Carbon monoxide	Inhalation	*Acute:* headache, lethargy; in extreme exposures, death through chemical asphyxiation

Table C.1 continued

Type of Stress or Agent	Major Route of Exposure	Potential Effects of Overexposure (Acute/Chronic)
Nitrogen, argon	Inhalation	*Acute:* Displacement of oxygen—potential asphyxiation
Sulfur dioxide	Inhalation	*Acute:* eye and respiratory irritation *Chronic:* bronchitis

Table C.2 Harmful Effects Resulting from Overexposure to Selected Physical Agents

Type of Stress or Agent	Potential Effects of Overexposure (Acute/Chronic)
Noise	*Acute:* headache, ringing ears *Chronic:* hearing loss
Radiation	
Ionizing	
(e.g., X-rays, alpha, beta, gamma radiation)	*Chronic:* cancer, leukemia, cataracts, sterility
Non-ionizing	
Ultraviolet (e.g., sun, weld arc, lasers)	*Acute:* dermatitis (skin erythema or sunburn), weld flash *Chronic:* skin cancer, cataracts
Infrared (e.g., ovens, molten metals, lasers)	*Acute:* tissue heating *Chronic:* cataracts
Microwave (e.g., ovens, heaters)	*Acute:* tissue heating, body temperature rise, localized tissue damage
Vibration	*Chronic:* circulatory disorders, Raynaud's Syndrome
Repetitive Motion	*Chronic:* tendinitis, carpal tunnel syndrome, epicondylitis, synovitis, muscular-skeletal problems, circulatory disorders

Appendix D

Threshold Limit Values and Biological Exposure Indices for 1989–1990

THRESHOLD LIMIT VALUES and BIOLOGICAL EXPOSURE INDICES for 1989–1990

American
Conference
of Governmental
Industrial Hygienists

New material or revisions for 1989–90* — *CHEMICAL SUBSTANCES:* 1) Addition of new substance TLVs for eight materials. 2) Revisions proposed for 17 listings including formaldehyde and all forms of nickel. 3) The year of most recent adoption of a TLV, or change thereto, is listed after the substance in the Adopted list. 4) For gases and vapors, TLVs, as converted to terms of mg/m³, have been recalculated and, where necessary, refined or corrected. A new section in the "Introduction" describes the basis for calculating the converted values. Since no TLV itself is changed, the recalculated converted values are not included in the "Notice of Intended Changes." *BIOLOGICAL EXPOSURE INDICES* are adopted for four listings and proposals are made for four more substances. *PHYSICAL AGENTS:* Change in a note for Hand–Arm (Segmental) Vibration and for the Radiofrequency/ Microwave Radiation lower end so as not to overlap with the Notice of Intent to Establish a TLV for Extremely Low Frequency (ELF) Electric Fields.

The limits listed in this book are intended for use in the practice of industrial hygiene as guidelines or recommendations in the control of potential health hazards and for no other use. These limits *are not* fine lines between safe and dangerous concentration and *should not* be used by anyone untrained in the discipline of industrial hygiene. It is imperative that the user of this book read the Introduction to each section before applying the recommendations contained herein as this is an integral part to the understanding and use of this book. The American Conference of Governmental Industrial Hygienists (ACGIH) disclaims liability with respect to the use of the TLVs.

Interpretation of TLVs: All requests for interpretations of TLVs must be in writing, in care of the Executive Secretary, ACGIH. Interpretations will be made only after full committee review and approval. Interpretations will be published annually in the *Supplemental Documentation.*

Documentation of the Threshold Limit Values and Biological Exposure Indices: A separate companion piece to the TLVs/BEIs is issued by ACGIH under this title. This publication gives the pertinent scientific information and data with reference to literature sources that were used to base each limit or index. For better understanding of the TLVs and BEIs, it is essential that the Documentation be consulted when the TLVs are being used or when invoking the BEIs. For further information, contact the Publications Office, ACGIH.

Requests for interpretations, reproduction permission, information, and/or placement of an order may use the following address: ACGIH, 6500 Glenway Ave., Bldg. D-7, Cincinnati, OH 45211-4438; telephone 513/661-7881; FAX 513/661-7195.

---

* Major text changes are indicated by a vertical rule in the margin.

The American Conference of Governmental Industrial Hygienists was organized in 1938 by a group of governmental industrial hygienists who desired a medium for the free exchange of ideas, experiences and the promotion of standards and techniques in industrial health. The Conference is a professional society, not an official Government Agency.

It is an organization devoted to the development of administrative and technical aspects of worker health protection. The association has contributed substantially to the development and improvement of official industrial health sevices to industry and labor. The committees on Industrial Ventilation and Threshold Limit Values are recognized throughout the world for their expertise and contributions to industrial hygiene.

Membership is limited to professional personnel in governmental agencies or educational institutions engaged in occupational safety and health programs. The more than 3700 members from across the United States and around the world give the organization an international scope.

ISBN: 0-936712-78-3

TABLE OF CONTENTS

Policy Statement on the Uses of TLVs and BEIs 60

Chemical Substances

Introduction 63
Adopted Threshold Limit Values 72
Notice of Intended Changes 109
Adopted Appendices
- A. Carcinogens 111
- B. Substances of Variable Composition 112
- C. Threshold Limit Values for Mixtures............. 113
- D. Particle Size-Selective Sampling Criteria for Airborne Particulate Matter 116
- E. Chemical Substances Under Study 118

Biological Exposure Indices

Introduction 123
Adopted Indices 128
Notice of Intent to Establish....................... 132
Chemical Substances Under Study to Establish BEIs ... 134

Physical Agents

Introduction 137
Adopted Threshold Limit Values
- Heat Stress 138
- Cold Stress 145
- Hand-Arm (Segmental) Vibration 157
- Ionizing Radiation 162
- Lasers 162
- Noise
  - Continuous 172
  - Impulsive or Impact 173
- Radiofrequency/Microwave Radiation 174
- Ultraviolet Radiation 178
- Airborne Upper Sonic and Ultrasonic Acoustic Radiation . 181

Notice of Intended Changes
- Ultraviolet Radiation 182

Notice of Intent to Establish
- Light and Near-Infrared Radiation 186
- Static Magnetic Fields........................... 190
- Extremely Low Frequency Electric Fields 190

Agents Under Study 191

POLICY STATEMENT ON THE USES OF TLVs AND BEIs

The Threshold Limit Values (TLVs) and Biological Exposure Indices (BEIs) are developed as guidelines to assist in the control of health hazards. These recommendations or guidelines are intended for use in the practice of industrial hygiene, to be interpreted and applied only by a person trained in this discipline. They are not developed for use as legal standards, and the American Conference of Governmental Industrial Hygienists (ACGIH) does not advocate their use as such. However, it is recognized that in certain circumstances individuals or organizations may wish to make use of these recommendations or guidelines as a supplement to their occupational safety and health program. The ACGIH will not oppose their use in this manner, if the use of TLVs and BEIs in these instances will contribute to the overall improvement in worker protection. However, the user must recognize the constraints and limitations subject to their proper use and bear the responsibility for such use.

The introductions to the TLV/BEI booklet and the TLV/BEI Documentation provide the philosophical and practical bases for the uses and limitations of the TLVs and BEIs. To extend those uses of the TLVs and BEIs to include other applications, such as use without the judgment of an industrial hygienist, application to a different population, development of new exposure/recovery time models, or new effect endpoints, stretch the reliability and even viability of the data base for the TLV or BEI as evidenced by the individual documentations.

It is not appropriate for individuals or organizations to impose on the TLVs or the BEIs their concepts of what the TLVs or BEIs should be or how they should be applied or to transfer regulatory standards requirements to the TLVs or BEIs.

The Policy Statement on the Uses of TLVs/BEIs was approved by the Board of Directors of ACGIH on March 1, 1988.

Threshold Limit Values for Chemical Substances in the Work Environment

Adopted by ACGIH with Intended Changes for 1989–90

1988-89 CHEMICAL SUBSTANCES TLV COMMITTEE

Ernest Mastromatteo, M.D., University of Toronto—*Chair*
James R. Crawl, CIH, US Navy
D. Dwight Culver, M.D., University of California-Irvine
Janet Fekete, MS, MIOSHA
Stanley Haimes, M.D., CIH, University of South Florida
John L.S. Hickey, Ph.D., CIH, Retired
William S. Lainhart, M.D., Retired
Trent R. Lewis, Ph.D., Retired
Jesse Lieberman, PE, CIH, Retired
Robert F. Phalen, Ph.D., University of California—Irvine, Air Sampling Procedures Liaison
Ronald S. Ratney, Ph.D., CIH, OSHA
Meier Schneider, PE, CIH, Retired
Raghubir Sharma, Ph.D., CIH, Utah State University
Judith A. Sparer, CIH, Yale University
Robert Spirtas, Dr. PH, National Institutes of Health
Vera F. Thomas, Ph.D., University of Miami, BEI Liaison
William D. Wagner, NIOSH
Elizabeth K. Weisburger, Ph.D., Retired, Corresponding Liaison
Calvin Willhite, Ph.D., State of Washington (4/89)
Margie E. Zalesak, CIH, USDL/MSHA

CONSULTANTS

Gerald L. Kennedy, Jr., Haskell Laboratory, E.I. du Pont de Nemours, Company, Inc.
Georg Kimmerle, M.D., Bayer AG, German MAK Commission Liaison
Theodore R. Torkelson, Sc.D., Retired
Richard A. Youngstrom, CIH, CSP, IUE Local 201

INTRODUCTION TO THE CHEMICAL SUBSTANCES*

Threshold Limit Values (TLVs) refer to airborne concentrations of substances and represent conditions under which it is believed that nearly all workers may be repeatedly exposed day after day without adverse effect. Because of wide variation in individual susceptibility, however, a small percentage of workers may experience discomfort from some substances at concentrations at or below the threshold limit; a smaller percentage may be affected more seriously by aggravation of a pre-existing condition or by development of an occupational illness. Smoking of tobacco is harmful for several reasons. Smoking may act to enhance the biological effects of chemicals encountered in the workplace and may reduce the body's defense mechanisms against toxic substances.

Individuals may also be hypersusceptible or otherwise unusually responsive to some industrial chemicals because of genetic factors, age, personal habits (smoking, alcohol, or other drugs), medication, or previous exposures. Such workers may not be adequately protected from adverse health effects from certain chemicals at concentrations at or below the threshold limits. An occupational physician should evaluate the extent to which such workers require additional protection.

TLVs are based on the best available information from industrial experience, from experimental human and animal studies, and, when possible, from a combination of the three. The basis on which the values are established may differ from substance to substance; protection against impairment of health may be a guiding factor for some, whereas reasonable freedom from irritation, narcosis, nuisance, or other forms of stress may form the basis for others.

The amount and nature of the information available for establishing a TLV varies from substance to substance; consequently, the precision of the estimated TLV is also subject to variation and the latest TLV *Documentation* should be consulted in order to assess the extent of the data available for a given substance.

These limits are intended for use in the practice of industrial hygiene as guidelines or recommendations in the control of potential health hazards and for no other use, e.g., in the evaluation or control of community air pollution nuisances; in estimating the toxic potential of continuous, uninterrupted exposures or other extended work periods; as proof or disproof of an existing disease or physical condition; or adoption by countries whose working conditions differ from those in the

---

*Major text changes are indicated by a vertical rule in the margin.

United States of America and where substances and processes differ. These limits *are not* fine lines between safe and dangerous concentration nor are they a relative index of toxicity. They *should not* be used by anyone untrained in the discipline of industrial hygiene.

The TLVs, as issued by the American Conference of Governmental Industrial Hygienists, are recommendations and should be used as guidelines for good practices. In spite of the fact that serious injury is not believed likely as a result of exposure to the threshold limit concentrations, the best practice is to maintain concentrations of all atmospheric contaminants as low as is practical.

The American Conference of Governmental Industrial Hygienists disclaims liability with respect to the use of TLVs.

Notice of Intent. At the beginning of each year, proposed actions of the Chemical Substances TLV Committee for the forthcoming year are issued in the form of a "Notice of Intended Changes." This Notice provides an opportunity for comment and *solicits suggestions of substances to be added to the list. The suggestions should be accompanied by substantiating evidence.* The "Notice of Intended Changes" is presented after the Adopted Values in this section (and in the other two sections) of the booklet. Values listed in parenthesis in the "Adopted" list are to be used during the period in which a proposed change for that Value is listed in the Notice of Intended Changes.

Definitions. Three categories of Threshold Limit Values (TLVs) are specified herein, as follows:

a) *Threshold Limit Value–Time-Weighted Average (TLV–TWA)*— the time-weighted average concentration for a normal 8-hour workday and a 40-hour workweek, to which nearly all workers may be repeatedly exposed, day after day, without adverse effect.

b) *Threshold Limit Value–Short-Term Exposure Limit (TLV–STEL)*—the concentration to which workers can be exposed continuously for a short period of time without suffering from 1) irritation, 2) chronic or irreversible tissue damage, or 3) narcosis of sufficient degree to increase the likelihood of accidental injury, impair self-rescue or materially reduce work efficiency, and provided that the daily TLV–TWA is not exceeded. It is not a separate independent exposure limit; rather, it supplements the time-weighted average (TWA) limit where there are recognized acute effects from a substance whose toxic effects are primarily of a chronic nature. STELs are recommended only where toxic effects have been reported from high short-term exposures in either humans or animals.

A STEL is defined as a 15-minute TWA exposure which should not be exceeded at any time during a work day even if the 8-hour TWA is within the TLV–TWA. Exposures above the TLV–TWA up to the STEL should not be longer than 15 minutes and should not occur more than four times per day. There should be at least 60 minutes between successive exposures in this range. An averaging period other than 15 minutes may be recommended when this is warranted by observed biological effects.

c) *Threshold Limit Value–Ceiling (TLV–C)*—the concentration that should not be exceeded during any part of the working exposure.

In conventional industrial hygiene practice if instantaneous monitoring is not feasible, then the TLV–C can be assessed by sampling over a 15-minute period except for those substances which may cause immediate irritation when exposures are short.

For some substances, e.g., irritant gases, only one category, the TLV–Ceiling, may be relevant. For other substances, one or two categories may be relevant, depending upon their physiologic action. It is important to observe that if any one of these types of TLVs is exceeded, a potential hazard from that substance is presumed to exist.

The Chemical Substances TLV Committee holds to the opinion that TLVs based on physical irritation should be considered no less binding than those based on physical impairment. There is increasing evidence that physical irritation may initiate, promote, or accelerate physical impairment through interaction with other chemical or biologic agents.

Time-Weighted Average(TWA) vs Ceiling (C) Limits. TWAs permit excursions above the TLV provided they are compensated by equivalent excursions below the TLV–TWA during the workday. In some instances, it may be permissible to calculate the average concentration for a workweek rather than for a workday. The relationship between the TLV and permissible excursion is a rule of thumb and in certain cases may not apply. The amount by which the TLVs may be exceeded for short periods without injury to health depends upon a number of factors such as the nature of the contaminant, whether very high concentrations—even for short periods—produce acute poisoning, whether the effects are cumulative, the frequency with which high concentrations occur, and the duration of such periods. All factors must be taken into consideration in arriving at a decision as to whether a hazardous condition exists.

Although the TWA concentration provides the most satisfactory, practical way of monitoring airborne agents for compliance with the TLVs, there are certain substances for which it is inappropriate. In the latter group are substances which are predominantly fast acting and whose TLV is more appropriately based on this particular response. Substances with this type of response are best controlled by a ceiling limit that should not be exceeded. It is implicit in these definitions that the manner of sampling to determine noncompliance with the limits for each group must differ; a single, brief sample, that is applicable to a ceiling limit, is not appropriate to the TWA; here, a sufficient number of samples are needed to permit a TWA concentration throughout a complete cycle of operations or throughout the workshift.

Whereas the ceiling limit places a definite boundary that concentrations should not be permitted to exceed, the TWA requires an explicit limit to the excursions that are permissible above the listed TLVs. It should be noted that the same factors are used by the Chemical Substances TLV Committee in determining the magnitude of the value of the STELs, or whether to include or exclude a substance for a ceiling listing.

Excursion Limits. For the vast majority of substances with a TLV–TWA, there is not enough toxicological data available to warrant a STEL. Nevertheless, excursions above the TLV–TWA should be controlled even where the 8-hour TLV–TWA is within recommended limits. Earlier editions of the TLV list included such limits whose values depended on the TLV–TWAs of the substance in question.

While no rigorous rationale was provided for these particular values, the basic concept was intuitive: in a well-controlled process exposure, excursions should be held within some reasonable limits. Unfortunately, neither toxicology nor collective industrial hygiene experience provide a solid basis for quantifying what those limits should be. The approach here is that the maximum recommended excursion should be related to variability generally observed in actual industrial processes. In reviewing large numbers of industrial hygiene surveys conducted by the National Institute for Occupational Safety and Health, Leidel, Busch, and Crouse[(1)] found that short-term exposure measurements were generally lognormally distributed with geometric standard deviations mostly in the range of 1.5 to 2.0.

While a complete discussion of the theory and properties of the lognormal distribution is beyond the scope of this section, a brief description of some important terms is presented. The measure of central tendency in a lognormal description is the antilog of the mean logarithm of the sample values. The distribution is skewed, and the geometric mean is always smaller than the arithmetic mean by an amount which depends on the geometric standard deviation. In the lognormal distribution, the geometric standard deviation (sd_g) is the antilog of the standard deviation of the sample value logarithms and 68.26% of all values lie between m_g/sd_g and $m_g \times sd_g$.

If the short-term exposure values in a given situation have a geometric standard deviation of 2.0, 5% of all values will exceed 3.13 times the geometric mean. If a process displays a variability greater than this, it is not under good control and efforts should be made to restore control. This concept is the basis for the following excursion limit recommendations which apply to TLV-TWAs that do not have STELs:

Excursions in worker exposure levels may exceed 3 times the TLV–TWA for no more than a total of 30 minutes during a workday, and under no circumstances should they exceed 5 times the TLV–TWA, provided that the TLV–TWA is not exceeded.

The approach is a considerable simplification of the idea of the lognormal concentration distribution but is considered more convenient to use by the practicing industrial hygienist. If exposure excursions are maintained within the recommended limits, the geometric standard deviation of the concentration measurements will be near 2.0 and the goal of the recommendations will be accomplished.

When the toxicological data for a specific substance are available to establish a STEL, this value takes precedence over the excursion limit regardless of whether it is more or less stringent.

"Skin" Notation. Listed substances followed by the designation "Skin" refer to the potential contribution to the overall exposure by the cutaneous route including mucous membranes and eye, either by airborne or, more particularly, by direct contact with the substance. Vehicles can alter skin absorption.

Little quantitative data are available describing absorption of vapors and gases through the skin. The rate of absorption is a function of the concentration to which the skin is exposed.

Substances having a skin notation and a low TLV may present a problem at high airborne concentrations, particularly if a significant area of the skin is exposed for a long period of time. Protection of the respiratory tract, while the rest of the body surface is exposed to a high concentration, may present such a situation.

Biological monitoring should be considered to determine the relative contribution of dermal exposure to the total dose.

This attention-calling designation is intended to suggest appropriate measures for the prevention of cutaneous absorption so that the TLV is not invalidated.

Mixtures. Special consideration should be given also to the application of the TLVs in assessing the health hazards which may be associated with exposure to mixtures of two or more substances. A brief discussion of basic considerations involved in developing TLVs for mixtures, and methods for their development, amplified by specific examples, are given in Appendix C.

Respirable and Total Dust. For solid substances and liquified mists, TLVs are expressed in terms of total dust, except where the term "respirable dust" is used. See Appendix D, Particle Size-Selective Sampling Criteria for Airborne Particulate Matter, for the definition of respirable dust.

Particulates Not Otherwise Classified (PNOC). In contrast to fibrogenic dusts which cause scar tissue to be formed in lungs when inhaled in excessive amounts, so-called "nuisance" dusts have a long history of little adverse effect on lungs and do not produce significant organic disease or toxic effect when exposures are kept under reasonable control. Such dusts have also been called (biologically) "inert" dusts, but the latter term is inappropriate to the extent that there is no dust which does not evoke some cellular response in the lung when inhaled in sufficient amount. However, the lung-tissue reaction caused by inhalation of nuisance dusts has the following characteristics: 1) the architecture of the air spaces remains intact; 2) collagen (scar tissue) is not formed to a significant extent; and 3) the tissue reaction is potentially reversible.

Excessive concentrations of PNOCs in the workroom air may seriously reduce visibility; may cause unpleasant deposits in the eyes, ears, and nasal passages (e.g., Portland cement dust); or cause injury to the skin or mucous membranes by chemical or mechanical action *per se* or by the rigorous skin cleansing procedures necessary for their removal.

A TLV–TWA of 10 mg/m^3 of total dust containing no asbestos and $< 1\%$ crystalline silica is recommended for substances in these categories and for which no specific TLVs have been assigned. This limit, for a normal workday, does not apply to brief exposures at higher concentrations. Neither does it apply to those substances which may cause physiologic impairment at lower concentrations but for which a TLV has not yet been adopted. The separate nuisance dust appendix was deleted in favor of individual entries in the adopted listing for those substances formerly carried in this appendix.

Simple Asphyxiants — 'Inert" Gases or Vapors. A number of gases and vapors, when present in high concentrations in air, act primarily as simple asphyxiants without other significant physiologic effects. A TLV may not be recommended for each simple asphyxiant because the limiting factor is the available oxygen. The minimal oxygen content should be 18% by volume under normal atmospheric pressure (equivalent to a partial pressure, pO_2 of 135 torr). Atmospheres deficient in O_2 do not provide adequate warning and most simple asphyxiants are odorless. Several simple asphyxiants present an explosion hazard. Account should be taken of this factor in limiting the concentration of the asphyxiant. The separate appendix, which contained examples of some asphyxiants, has been deleted. Those substances formerly carried in that appendix continue as alphabetical entries in the Adopted listing.

Physical Factors. It is recognized that such physical factors as heat, ultraviolet and ionizing radiation, humidity, abnormal pressure (altitude), and the like may place added stress on the body so that the effects from exposure at a TLV may be altered. Most of these stresses act adversely to increase the toxic response of a substance. *Although most TLVs have built-in safety factors to guard against adverse effects to moderate deviations* from normal environments, the safety factors of most substances are not of such a magnitude as to take care of gross deviations. For example, continuous work at temperatures above 90°F, or overtime extending the workweek more than 25%, might be considered gross deviations. In such instances, judgment must be exercised in the proper adjustments of the TLVs.

Unlisted Substances. Many substances present or handled in industrial processes do not appear on the TLV list. In a number of instances, the material is rarely present as a particulate, vapor, or other airborne contaminant, and a TLV is not necessary. In other cases, sufficient information to warrant development of a TLV, even on a tentative basis, is not available to the Chemical Substances TLV Committee. Other substances, of low toxicity, could be classified as Particulates Not Otherwise Classified.

In addition, there are some substances of not inconsiderable toxicity that have been omitted primarily because only a limited number of workers (e.g., employees of a single plant) are known to have potential exposure to possibly harmful concentrations.

Unusual Work Schedules. Application of TLVs to workers on work schedules markedly different from the conventional 8-hour day, 40-hour week requires particular judgement in order to provide, for such workers, protection equal to that provided to workers on conventional work shifts.

As tentative guidance, field hygienists are referred to the "Brief and Scala model" and the "OSHA model," both of which are described and explained at length in Patty.[2]

The Brief and Scala model reduces the TLV proportionately for both increased exposure time and reduced recovery (nonexposure) time, and it is the more conservative of the two models. The OSHA model classifies air contaminants by type of toxic effect and recommends different adjustment procedures for the exposure limits (including zero adjustment) on that basis.

The two models are generally intended to apply to work schedules longer than 8 hours/day or 40 hours/week. The models should not be used to justify very high exposures as "allowable" where the exposure periods are short (e.g., exposure to 8 times the TLV–TWA for one hour and zero exposure during the remainder of the shift). In this respect, the general limitations on TLV excursions and STELs should be applied to avoid inappropriate use of the models with very short exposure periods or shifts.

Since adjusted TLVs do not have the benefit of historical use and long-time observation, medical supervision during initial use of adjusted TLVs is advised. In addition, the hygienist should avoid unnecessary exposure of workers even if a model shows such exposures to be "allowable" and should not use models to justify higher-than-necessary exposures.

Both the OSHA and the Brief and Scala models are easier to use than some of the more complex models based on pharmacokinetic actions. However, hygienists thoroughly familiar with such models may find them more appropriate in specific instances. Use of such models usually requires knowledge of the biological half-life of each substance, and some models require additional data.

Short workweeks can allow workers to have two full-time jobs, perhaps with similar exposures, and may result in overexposure even if neither job by itself entails overexposure. Hygienists should be alert to such situations.

Conversion of TLVs in ppm to mg/m³. TLVs for gases and vapors are usually established in terms of parts per million of substance in air by volume (ppm). For convenience to the user, these TLVs are also listed here in terms of milligrams of substance per cubic meter of air (mg/m³). The conversion is based on 760 torr barometric pressure and 25°C (77°F), giving a conversion equation of:

$$\text{TLV in mg/m}^3 = \frac{(\text{TLV in ppm})\ (\text{gram molecular weight of the substance})}{24.45}$$

Resulting values are rounded to two significant figures below 100 and to three significant figures above 100. This is not done to give any converted value a greater precision than that of the original TLV, but to avoid increasing or decreasing the TLV significantly merely by the conversion of units.

The above equation may be used to convert TLVs to any degree of precision desired. When converting TLVs to mg/m³ units for other temperatures and pressures, the reference TLVs should be used as a starting point. These are the values expressed in ppm, except for benomyl, caprolactam, and sodium azide, which are established in terms of mg/m³. Also, when converting values expressed as an element (e.g., as Fe, as Ni), the molecular value of the element should be used, not that of the entire compound.

In making conversions for substances with variable molecular weights, appropriate molecular weights have been estimated or assumed (see the TLV *Documentation).*

Operational Guidelines. The ACGIH Board of Directors has adopted operational guidelines for the Chemical Substances TLV Committee. These guidelines prescribe: charge, authority, policies, membership, organization, and operating procedures. The policies include the appeals procedures. Copies of the guidelines document are available from the ACGIH Publications Office at a cost of $5 per copy. The guidelines were also published in the March 1988 issue of *Applied Industrial Hygiene*. These guidelines permit public input.

References

1. Leidel, N.A.; Busch, K.A.; Crouse, W.E.: Exposure Measurement, Action Level and Occupational Environmental Variability. DHEW (NIOSH) Pub. No. 76-131 (December 1975).
2. Paustenbach, D.J.: Occupational Exposure Limits, Pharmacokinetics, and Unusual Work Schedules. In: Patty's Industrial Hygiene and Toxicology, 2nd ed., Vol. 3A, The Work Environment, Chap. 6, pp. 111–277. L.J. Cralley and L.V. Cralley, Eds. John Wiley and Sons, Inc., New York (1985).

NOTE: In the interest of keeping this book as compact as possible, the footnotes for the adopted values appear on right-hand pages only.

Substance [CAS #]	TWA ppm[a)]	TWA mg/m³[b)]	STEL ppm[a)]	STEL mg/m³[b)]
	ADOPTED VALUES			
▪Acetaldehyde [75-07-0] (1976) .	100	180	150	270
Acetic acid [64-19-7] (1976) . .	10	25	15	37
Acetic anhydride [108-24-7] (1976)	C 5	C 21	—	—
•Acetone [67-64-1] (1982)	750	1780	1000	2380
•Acetonitrile [75-05-8]—Skin (1976)	40	67	60	101
Acetylene [74-86-2] (1981) . . .	—(d)	—	—	—
Acetylene dichloride, *see* 1,2-Dichloroethylene				
Acetylene tetrabromide [79-27-6] (1986)	1	14	—	—
Acetylsalicylic acid (Aspirin) [50-78-2] (1980)	—	5	—	—
Acrolein [107-02-8] (1976)	0.1	0.23	0.3	0.69
▪Acrylamide [79-06-1]—Skin (1987)	—	0.03,A2	—	—
‡Acrylic acid [79-10-7]	(10)	(29)	—	—
•▪Acrylonitrile [107-13-1]—Skin (1984)	2,A2	4.3,A2	—	—
•▪Aldrin [309-00-2]—Skin (1986)	—	0.25	—	—
Allyl alcohol [107-18-6]—Skin (1976)	2	4.8	4	9.5
▪Allyl chloride [107-5-1] (1976) .	1	3.0	2	6.0
Allyl glycidyl ether (AGE) [106-92-3]—Skin (1976) . . .	5	23	10	47
Allyl propyl disulfide [2179-59-1] (1976)	2	12	3	18
α-Alumina, *see* Aluminum oxide				
Aluminum [7429-90-5], as Al				
Metal dust (1986)	—	10	—	—
Pyro powders (1979)	—	5	—	—
Welding fumes (1979)	—	5	—	—
Soluble salts (1979)	—	2	—	—
Alkyls (NOC†) (1979)	—	2	—	—
Aluminum oxide [1344-28-1], as Al (1986)	—	10(e)	—	—
▪4-Aminodiphenyl [92-67-1]—Skin (1972)	—	A1	—	—
2-Aminoethanol, *see* Ethanolamine				
2-Aminopyridine [504-29-0] (1986)	0.5	2.0	—	—
3-Amino 1,2,4-triazole, *see* Amitrole				

		ADOPTED VALUES			
		TWA		STEL	
Substance	[CAS #]	ppm[a)]	mg/m$^{3b)}$	ppm[a)]	mg/m$^{3b)}$
■Amitrole [61-82-5] (1986)		—	0.2	—	—
Ammonia [7664-41-7] (1976)		25	17	35	24
Ammonium chloride fume [12125-02-9] (1976)		—	10	—	20
Ammonium perfluorooctanoate [3825-26-1] (1988)		—	0.1	—	—
Ammonium sulfamate [7773-06-0] (1986)		—	10	—	—
Amosite, *see* Asbestos					
n-Amyl acetate [628-63-7] (1987)		100	532	—	—
sec-Amyl acetate [626-38-0] (1987)		125	665	—	—
◄■Aniline [62-53-3] & homologues—Skin (1986)		2	7.6	—	—
■Anisidine [29191-52-4] (o-, p-isomers)—Skin (1977)		0.1	0.50	—	—
Antimony [7440-36-0] & compounds, as Sb (1980)		—	0.5	—	—
■Antimony trioxide [1309-64-4]					
Handling and use, as Sb (1978)		—	0.5	—	—
Production (1980)		—	A2	—	—
ANTU [86-88-4] (1986)		—	0.3	—	—
Argon [7440-37-1] (1981)		—(d)	—	—	—

(a) Parts of vapor or gas per million parts of contaminated air by volume at 25°C and 760 torr.
(b) Milligrams of substance per cubic meter of air. When entry is in this column only, the value is exact; when listed with a ppm entry, it is approximate.
(d) Simple asphyxiant; see definition in the "Introduction to the Chemical Substances."
(e) The value is for total dust containing no asbestos and < 1% crystalline silica.
Capital letters A & B refer to Appendices; C denotes ceiling limit.
( ) Adopted values enclosed are on the Notice of Intended Changes list. Consult it for current proposal for change.
◄ Identifies substances for which there are also BEIs (*see* BEI section). Substances identified in the BEI documentations for methemoglobin inducers (for which methemoglobin is the principle toxicity) and organophosphorus cholinesterase inhibitors are part of this notation.
■ Substance identified by other sources as a suspected or confirmed human carcinogen. See the compliation in the Appendix to the Documentation of TLVs, pp. A-5(86)–A-9(86).
• Substance for which OSHA and/or NIOSH has a Permissible Exposure Limit (PEL) or a Recommended Exposure Limit (REL) lower than the TLV.
‡ See Notice of Intended Changes.
† NOC = not otherwise classified.

Substance [CAS #]	TWA ppm[a)]	TWA mg/m³[b)]	STEL ppm[a)]	STEL mg/m³[b)]
	ADOPTED VALUES			
•▪Arsenic [7440-38-2] & soluble compounds, as As (1980)	—	0.2	—	—
Arsenic trioxide production [1327-53-3] (1980)	—	A2	—	—
•Arsine [7784-42-1] (1977)	0.05	0.16	—	—
•▪Asbestos[(f)]				
• Amosite [12172-73-5] (1980)		0.5 fiber/cc, A1		
• Chrysotile [12001-29-5] (1980)		2 fibers/cc, A1		
• Crocidolite [12001-28-4] (1980)		0.2 fiber/cc, A1		
• Other forms (1980)		2 fibers/cc, A1		
Asphalt (petroleum) fumes [8052-42-4] (1987)	—	5	—	—
Atrazine [1912-24-9] (1983) . . .	—	5	—	—
◂Azinphos-methyl [86-50-0]—Skin (1986)	—	0.2	—	—
Barium [7440-39-3], soluble compounds, as Ba (1977)	—	0.5	—	—
Barium sulfate [7727-43-7] (1986).	—	10[(e)]	—	—
Benomyl [17804-35-2] (1986) . .	0.84	10	—	—
◂▪•Benzene [71-43-2] (1987)	10,A2	32,A2	—	—
▪Benzidine [92-87-5]—Skin (1982)	—	A1	—	—
p-Benzoquinone, *see* Quinone				
Benzoyl peroxide [94-36-0] (1977).	—	5	—	—
▪Benzo(a)pyrene [50-32-8] (1976)	—	A2	—	—
▪Benzyl chloride [100-44-7] (1977)	1	5.2	—	—
•▪Beryllium and compounds, as Be [7440-41-7] (1979) . . .	—	0.002,A2	—	—
Biphenyl [92-52-4] (1987)	0.2	1.3	—	—
Bismuth telluride [1304-82-1] (1986)	—	10	—	—
Se-doped (1986)	—	5	—	—
Borates, tetra, sodium salts [1303-96-4]				
Anhydrous (1977)	—	1	—	—
Decahydrate (1977)	—	5	—	—
Pentahydrate (1977)	—	1	—	—
Boron oxide [1303-86-2] (1986)	—	10	—	—

		ADOPTED VALUES			
		TWA		STEL	
Substance	[CAS #]	ppm[a)]	mg/m³[b)]	ppm[a)]	mg/m³[b)]
Boron tribromide [10294-33-4] (1986)		C 1	C 10	—	—
Boron trifluoride [7637-07-2] (1977)		C 1	C 2.8	—	—
Bromacil [314-40-9] (1986)		1	11	—	—
Bromine [7726-95-6] (1976)		0.1	0.66	0.3	2.0
Bromine pentafluoride [7789-30-2] (1986)		0.1	0.72	—	—
Bromochloromethane, *see* Chlorobromomethane					
Bromoform [75-25-2] — Skin (1977)		0.5	5.2	—	—
▪1,3-Butadiene [106-99-0] (1986)		10,A2	22,A2	—	—
Butane [106-97-8] (1981)		800	1900	—	—
Butanethiol, *see* Butyl mercaptan					
2-Butanone, *see* Methyl ethyl ketone (MEK)					
2-Butoxyethanol [111-76-2] — Skin (1987)		25	121	—	—
n-Butyl acetate [123-86-4] (1976)		150	713	200	950
sec-Butyl acetate [105-46-4] (1987)		200	950	—	—
tert-Butyl acetate [540-88-5] (1987)		200	950	—	—
Butyl acrylate [141-32-2] (1978)		10	52	—	—
n-Butyl alcohol [71-36-3] — Skin (1977)		C 50	C 152	—	—
‡sec-Butyl alcohol [78-92-2]		100	303	(150)	(455)
tert-Butyl alcohol [75-65-0] (1976)		100	303	150	455

(e) The value is for total dust containing no asbestos and < 1% crystalline silica.

(f) Fibers longer than 5 μm and with an aspect ratio equal to or greater than 3:1 as determined by the membrane filter method at 400-450X magnification (4 mm objective) phase contrast illumination.

Capital letters A & B refer to Appendices; C denotes ceiling limit.

( ) Adopted values enclosed are on the Notice of Intended Changes list. Consult it for current proposal for change.

◄ Identifies substances for which there are also BEIs (*see* BEI section). Substances identified in the BEI documentations for methemoglobin inducers (for which methemoglobin is the principle toxicity) and organophosphorus cholinesterase inhibitors are part of this notation.

▪ Substance identified by other sources as a suspected or confirmed human carcinogen. See the compliation in the Appendix to the Documentation of TLVs, pp. A-5(86)–A-9(86).

• Substance for which OSHA and/or NIOSH has a Permissible Exposure Limit (PEL) or a Recommended Exposure Limit (REL) lower than the TLV.

‡ See Notice of Intended Changes.

Substance [CAS #]	TWA ppm[a]	TWA mg/m^3[b]	STEL ppm[a]	STEL mg/m^3[b]
	ADOPTED VALUES			
Butylamine [109-73-9]—Skin (1976)	C 5	C 15	—	—
•tert-Butyl chromate, as CrO_3 [1189-85-1]—Skin (1977)	—	C 0.1	—	—
•n-Butyl glycidyl ether (BGE) [2426-08-6] (1981)	25	133	—	—
n-Butyl lactate [138-22-7] (1977)	5	30	—	—
Butyl mercaptan [109-79-5] (1977)	0.5	1.8	—	—
o-sec-Butylphenol [89-72-5]—Skin (1980)	5	31	—	—
p-tert-Butyltoluene [98-51-1] (1976)	10	61	20	122
‡▪◂Cadmium [7440-43-9] dusts & salts, as Cd	—	(0.05)	—	—
‡Cadmium oxide [1306-19-0]				
Fume, as Cd	—	(C 0.05)	—	—
Production	—	(0.05)	—	—
Calcium carbonate [1317-65-3] (1986)	—	10[e]	—	—
Calcium cyanamide [156-62-7] (1986)	—	0.5	—	—
Calcium hydroxide [1305-62-0] (1978)	—	5	—	—
Calcium oxide [1305-78-8] (1978)	—	2	—	—
Calcium silicate [1344-95-2] (1974)	—	10[e]	—	—
Calcium sulfate [7778-18-9] (1986)	—	10[e]	—	—
Camphor, synthetic [76-22-2] (1976)	2	12	3	19
‡Caprolactam [105-60-2]				
Dust	—	(1)	—	(3)
Vapor	(4.3)	(20)	(8.6)	(40)
Captafol [2425-06-1]—Skin (1977)	—	0.1	—	—
Captan [133-06-2] (1986)	—	5	—	—
Carbaryl [63-25-2] (1986)	—	5	—	—
Carbofuran [1563-66-2] (1977)	—	0.1	—	—
▪Carbon black [1333-86-4] (1986)	—	3.5	—	—
Carbon dioxide [124-38-9] (1986)	5000	9000	30,000	54,000

Substance [CAS #]	ADOPTED VALUES TWA ppm[a)]	TWA mg/m[3b)]	STEL ppm[a)]	STEL mg/m[3b)]
•◂Carbon disulfide [75-15-0] — Skin (1986)	10	31	—	—
•◂Carbon monoxide [630-08-0] (1976)	50	57	400	458
Carbon tetrabromide [558-13-4] (1976)	0.1	1.4	0.3	4.1
•■Carbon tetrachloride [56-23-5] — Skin (1986)	5,A2	31,A2	—	—
Carbonyl chloride, *see* Phosgene				
Carbonyl fluoride [353-50-4] (1981)	2	5.4	5	13
Catechol [120-80-9] (1977)	5	23	—	—
Cellulose (paper fiber) [9004-34-6] (1986)	—	10(e)	—	—
Cesium hydroxide [21351-79-1] (1977)	—	2	—	—
‡■Chlordane [57-74-9] — Skin	—	0.5	—	(2)
■Chlorinated camphene [8001-35-2] — Skin (1976)	—	0.5	—	1
‡Chlorinated diphenyl oxide [55720-99-5]	—	0.5	—	(2)
*Chlorine [7782-50-5] (1989)	0.5	1.5	1	3.0
Chlorine dioxide [10049-04-4] (1976)	0.1	0.28	0.3	0.83
Chlorine trifluoride [7790-91-2] (1977)	C 0.1	C 0.38	—	—

(e) The value is for total dust containing no asbestos and < 1% crystalline silica. Capital letters A & B refer to Appendices; C denotes ceiling limit.

( ) Adopted values enclosed are on the Notice of Intended Changes list. Consult it for current proposal for change.

◂ Identifies substances for which there are also BEIs (*see* BEI section). Substances identified in the BEI documentations for methemoglobin inducers (for which methemoglobin is the principle toxicity) and organophosphorus cholinesterase inhibitors are part of this notation.

■ Substance identified by other sources as a suspected or confirmed human carcinogen. See the compliation in the Appendix to the Documentation of TLVs, pp. A-5(86)–A-9(86).

• Substance for which OSHA and/or NIOSH has a Permissible Exposure Limit (PEL) or a Recommended Exposure Limit (REL) lower than the TLV.

‡ See Notice of Intended Changes.

* 1989–1990 Adoption.

Substance [CAS #]	TWA ppm[a)]	TWA mg/m^{3}[b)]	STEL ppm[a)]	STEL mg/m^{3}[b)]
Chloroacetaldehyde [107-20-0] (1977)	C 1	C 3.2	—	—
*Chloroacetone [78-95-5]—Skin (1989)	C 1	C 3.8	—	—
α-Chloroacetophenone [532-27-4] (1977)	0.05	0.32	—	—
‡Chloroacetyl chloride [79-04-9]	(0.05)	(0.23)	—	—
‡Chlorobenzene [108-90-7]	(75)	(345)	—	—
o-Chlorobenzylidene malononitrile [2698-41-1]—Skin (1983)	C 0.05	C 0.39	—	—
‡Chlorobromomethane [74-97-5]	200	1058	(250)	(1320)
2-Chloro-1,3-butadiene, *see* β-Chloroprene				
‡▪Chlorodifluoromethane [75-45-6]	1000	3540	(1250)	(4430)
‡•Chlorodiphenyl (42% chlorine) [53469-21-9]—Skin	—	1	—	(2)
‡•▪Chlorodiphenyl (54% chlorine) [11097-69-1]—Skin	—	0.5	—	(1)
1-Chloro,2,3-epoxy-propane, *see* Epichlorohydrin				
2-Chloroethanol, *see* Ethylene chlorohydrin				
Chloroethylene, *see* Vinyl chloride				
•▪Chloroform [67-66-3] (1986)	10,A2	49,A2	—	—
▪bis(Chloromethyl) ether [542-88-1] (1981)	0.001, A1	0.0047, A1	—	—
▪Chloromethyl methyl ether [107-30-2] (1983)	A2	A2	—	—
1-Chloro-1-nitropropane [600-25-9] (1981)	2	10	—	—
Chloropentafluoroethane [76-15-3] (1981)	1000	6320	—	—
‡Chloropicrin [76-06-2]	0.1	0.67	(0.3)	(2.0)
•▪β-Chloroprene [126-99-8]—Skin (1980)	10	36	—	—
o-Chlorostyrene [2039-87-4] (1976)	50	283	75	425
‡o-Chlorotoluene [95-49-8]	50	259	(75)	(388)
2-Chloro-6-(trichloromethyl) pyridine, *see* Nitrapyrin				

Substance [CAS #]	TWA ppm[a)]	TWA mg/m³[b)]	STEL ppm[a)]	STEL mg/m³[b)]
‡Chlorpyrifos [2921-88-2] — Skin	—	0.2	—	(0.6)
•Chromite ore processing (Chromate), as Cr (1978)	—	0.05,A1	—	—
Chromium [7440-47-3] Metal (1981)	—	0.5	—	—
Chromium (II) compounds, as Cr (1981)	—	0.5	—	—
Chromium (III) compounds, as Cr (1981)	—	0.5	—	—
•■◄Chromium (VI) compounds, as Cr				
•◄Water soluble (1981)	—	0.05	—	—
• Certain water insoluble (1981)	—	0.05,A1	—	—
•■Chromyl chloride [14977-61-8] (1982)	0.025	0.16	—	—
■Chrysene [218-01-9] (1981)	A2	A2	—	—
Chrysotile, *see* Asbestos				
‡Clopidol [2971-90-6]	—	10	—	(20)
Coal dust (1987)	—	2,(g) Respirable fraction		

(g) The value is for dust containing < 5% silica. For dust containing more than this percentage of free silica, the environment should be elevated against the TLV-TWA of 0.1 mg/m³ for respirable quartz. The concentration of respirable dust for the application of this limit is to be determined from the fraction passing a size-selector with the characteristics defined in the "c." paragraphs of Appendix D.

Capital letters A & B refer to Appendices; C denotes ceiling limit.

( ) Adopted values enclosed are on the Notice of Intended Changes list. Consult it for current proposal for change.

◄ Identifies substances for which there are also BEIs (*see* BEI section). Substances identified in the BEI documentations for methemoglobin inducers (for which methemoglobin is the principle toxicity) and organophosphorus cholinesterase inhibitors are part of this notation.

■ Substance identified by other sources as a suspected or confirmed human carcinogen. See the compliation in the Appendix to the Documentation of TLVs, pp. A-5(86)–A-9(86).

• Substance for which OSHA and/or NIOSH has a Permissible Exposure Limit (PEL) or a Recommended Exposure Limit (REL) lower than the TLV.

‡ See Notice of Intended Changes.

* 1989–1990 Adoption.

Substance [CAS #]	TWA ppm[a)]	TWA mg/m³[b)]	STEL ppm[a)]	STEL mg/m³[b)]
	ADOPTED VALUES			
•▪Coal tar pitch volatiles [65996-93-2], as benzene solubles (1981) . .	—	0.2,A1	—	—
▪Cobalt [7440-48-4], as Co				
Metal dust & fume (1987)	—	0.05	—	—
Cobalt carbonyl [10210-68-1], as Co (1983) .	—	0.1	—	—
Cobalt hydrocarbonyl [16842-03-8], as Co (1983) .	—	0.1	—	—
Copper [7440-50-8]				
Fume (1977)	—	0.2	—	—
Dusts & mists, as Cu (1986)	—	1	—	—
Cotton dust, raw (1986)	—	0.2[(h)]	—	—
•Cresol [1319-77-3], all isomers—Skin (1977)	5	22	—	—
Cristobalite, *see* Silica—Crystaline				
Crocidolite, *see* Asbestos				
▪Crotonaldehyde [4170-30-3] (1987)	2	5.7	—	—
‡Crufomate [299-86-5]	—	5	—	(20)
Cumene [98-82-8]—Skin (1987)	50	246	—	—
Cyanamide [420-04-2] (1977) .	—	2	—	—
Cyanides [151-50-8; 143-33-9], as CN—Skin (1977)	—	5	—	—
Cyanogen [460-19-5] (1977) . .	10	21	—	—
Cyanogen chloride [506-77-4] (1980)	C 0.3	C 0.75	—	—
Cyclohexane [110-82-7] (1987)	300	1030	—	—
Cyclohexanol [108-93-0]—Skin (1977)	50	206	—	—
Cyclohexanone [108-94-1]—Skin (1987)	25	100	—	—
Cyclohexene [110-83-8] (1977)	300	1010	—	—
Cyclohexylamine [108-91-8] (1977)	10	41	—	—
‡Cyclonite [121-82-4]—Skin . .	—	1.5	—	(3)
Cyclopentadiene [542-92-7] (1987)	75	203	—	—
Cyclopentane [287-92-3] (1987)	600	1720	—	—
Cyhexatin [13121-70-5] (1986) .	—	5	—	—
2,4-D [94-75-7] (1986)	—	10	—	—

Substance	[CAS #]	TWA ppm[a)]	TWA mg/m³[b)]	STEL ppm[a)]	STEL mg/m³[b)]
		ADOPTED VALUES			
■•DDT (Dichlorodiphenyl-trichloroethane)	[50-29-3] (1986)	—	1	—	—
Decaborane	[17702-41-9] — Skin (1976)	0.05	0.25	0.15	0.75
◄Demeton	[8065-48-3] — Skin (1986)	0.01	0.11	—	—
Diacetone alcohol	[123-42-2] (1987)	50	238	—	—
1,2-Diaminoethane, *see* Ethylenediamine					
Diatomaceous earth, *see* Silica—Amorphous					
◄Diazinon	[333-41-5] — Skin (1986)	—	0.1	—	—
■Diazomethane	[334-88-3] (1977)	0.2	0.34	—	—
Diborane	[19287-45-7] (1977)	0.1	0.11	—	—
1,2-Dibromoethane, *see* Ethylene dibromide					
2-N-Dibutylaminoethanol	[102-81-8] — Skin (1986)	2	14	—	—
Dibutyl phosphate	[107-66-4] (1976)	1	8.6	2	17
Dibutyl phthalate	[84-74-2] (1987)	—	5	—	—
■Dichloroacetylene	[7572-29-4] (1977)	C 0.1	C 0.39	—	—

(h) Lint-free dust as measured by the vertical elutriator cotton-dust sampler described in the *Transactions of the National Conference on Cotton Dust,* p. 33, by J.R. Lynch (May 2, 1970).

Capital letters A & B refer to Appendices; C denotes ceiling limit.

( ) Adopted values enclosed are on the Notice of Intended Changes list. Consult it for current proposal for change.

◄ Identifies substances for which there are also BEIs (*see* BEI section). Substances identified in the BEI documentations for methemoglobin inducers (for which methemoglobin is the principle toxicity) and organophosphorus cholinesterase inhibitors are part of this notation.

■ Substance identified by other sources as a suspected or confirmed human carcinogen. See the compliation in the Appendix to the Documentation of TLVs, pp. A-5(86)–A-9(86).

• Substance for which OSHA and/or NIOSH has a Permissible Exposure Limit (PEL) or a Recommended Exposure Limit (REL) lower than the TLV.

‡ See Notice of Intended Changes.

Substance	[CAS #]	TWA ppm[a)]	TWA mg/m³[b)]	STEL ppm[a)]	STEL mg/m³[b)]
		ADOPTED VALUES			
o-Dichlorobenzene	[95-50-1] — Skin (1977)	C 50	C 301	—	—
p-Dichlorobenzene	[106-46-7] (1977)	75	451	110	661
▪3,3′-Dichlorobenzidine	[91-94-1] — Skin (1976)	—	A2	—	—
Dichlorodifluoromethane	[75-71-8] (1986)	1000	4950	—	—
1,3-Dichloro-5,5-dimethyl hydantoin	[118-52-5] (1976)	—	0.2	—	0.4
•1,1-Dichloroethane	[75-34-3] (1976)	200	810	250	1010
1,2-Dichloroethane, *see* Ethylene dichloride					
1,1-Dichloroethylene, *see* Vinylidene chloride					
1,2-Dichloroethylene	[540-59-0] (1987)	200	793	—	—
Dichloroethyl ether	[111-44-4] — Skin (1976)	5	29	10	58
Dichlorofluoromethane	[75-43-4] (1980)	10	42	—	—
Dichloromethane, *see* Methylene chloride					
1,1-Dichloro-1-nitroethane	[594-72-9] (1986)	2	12	—	—
1,2-Dichloropropane, *see* Propylene dichloride					
▪Dichloropropene	[542-75-6] — Skin (1986)	1	4.5	—	—
2,2-Dichloropropionic acid	[75-99-0] (1980)	1	5.8	—	—
Dichlorotetrafluoroethane	[76-14-2] (1986)	1000	6990	—	—
◂Dichlorvos	[62-73-7] — Skin (1986)	0.1	0.90	—	—
Dicrotophos	[141-66-2] — Skin (1977)	—	0.25	—	—
Dicyclopentadiene	[77-73-6] (1977)	5	27	—	—
Dicyclopentadienyl iron	[102-54-5] (1986)	—	10	—	—
▪•Dieldrin	[60-57-1] — Skin (1986)	—	0.25	—	—
Diethanolamine	[111-42-2] (1980)	3	13	—	—
Diethylamine	[109-89-7] (1981)	10	30	25	75

Substance [CAS #]	TWA ppm[a)]	TWA mg/m³[b)]	STEL ppm[a)]	STEL mg/m³[b)]
2-Diethylaminoethanol [100-37-8] — Skin (1977) . . .	10	48	—	—
Diethylene triamine [111-40-0] — Skin (1977) . . .	1	4.2	—	—
Diethyl ether, *see* Ethyl ether				
Di(2-ethylhexyl)phthalate, *see* Di-sec-octyl phthalate				
Diethyl ketone [96-22-0] (1981)	200	705	—	—
Diethyl phthalate [84-66-2] (1987)	—	5	—	—
Difluorodibromomethane [75-61-6] (1986)	100	858	—	—
▪Diglycidyl ether (DGE) [2238-07-5] (1981)	0.1	0.53	—	—
Dihydroxybenzene, *see* Hydroquinone				
Diisobutyl ketone [108-83-8] (1977)	25	145	—	—
Diisopropylamine [108-18-9] — Skin (1977) . . .	5	21	—	—
Dimethoxymethane, *see* Methylal				
Dimethyl acetamide [127-19-5] — Skin (1986) . . .	10	36	—	—
Dimethylamine [124-40-3] (1977)	10	18	—	—
Dimethylaminobenzene, *see* Xylidene				
◂Dimethylaniline [121-69-7] (N,N-Dimethylaniline) — Skin (1976)	5	25	10	50
Dimethylbenzene, *see* Xylene				
▪Dimethyl carbamoyl chloride [79-44-7] (1978)	A2	A2	—	—
Dimethyl-1,2-dibromo-2,2-dichloroethyl phosphate, *see* Naled				
◂Dimethylformamide [68-12-2] — Skin (1986)	10	30	—	—

Capital letters A & B refer to Appendices; C denotes ceiling limit.

◂ Identifies substances for which there are also BEIs (*see* BEI section). Substances identified in the BEI documentations for methemoglobin inducers (for which methemoglobin is the principle toxicity) and organophosphorus cholinesterase inhibitors are part of this notation.

▪ Substance identified by other sources as a suspected or confirmed human carcinogen. See the compliation in the Appendix to the Documentation of TLVs, pp. A-5(86)–A-9(86).

• Substance for which OSHA and/or NIOSH has a Permissible Exposure Limit (PEL) or a Recommended Exposure Limit (REL) lower than the TLV.

		ADOPTED VALUES			
		TWA		STEL	
Substance	[CAS #]	ppm[a)]	mg/m³[b)]	ppm[a)]	mg/m³[b)]
2,6-Dimethyl-4-heptanone, *see* Diisobutyl ketone					
‡•▪1,1-Dimethylhydrazine	[57-14-7] — Skin	(0.5,A2)	(1.2,A2)	—	—
Dimethylnitrosoamine, *see* N-Nitrosodimethylamine					
Dimethylphthalate	[131-11-3] (1986)	—	5	—	—
▪Dimethyl sulfate	[77-78-1] — Skin (1977)	0.1,A2	0.52,A2	—	—
Dinitolmide	[148-01-6] (1976)	—	5	—	—
◂Dinitrobenzene	[528-29-0; 99-65-0; 100-25-4] (all isomers) — Skin (1986)	0.15	1.0	—	—
Dinitro-o-cresol	[534-52-1] — Skin (1986)	—	0.2	—	—
3,5-Dinitro-o-toluamide, *see* Dinitolmide					
•▪◂Dinitrotoluene	[25321-14-6] — Skin (1986)	—	1.5	—	—
•▪Dioxane	[123-91-1] — Skin (1986)	25	90	—	—
◂Dioxathion	[78-34-2] — Skin (1977)	—	0.2	—	—
Diphenyl, *see* Biphenyl					
Diphenylamine	[122-39-4] (1986)	—	10	—	—
Diphenylmethane diisocyanate, *see* Methylene bisphenyl isocyanate					
Dipropylene glycol methyl ether	[34590-94-8] — Skin (1976)	100	606	150	909
Dipropyl ketone	[123-19-3] (1981)	50	233	—	—
Diquat	[85-00-7] (1986)	—	0.5	—	—
▪•Di-sec-octyl phthalate	[117-81-7] (1976)	—	5	—	10
Disulfiram	[97-77-8] (1986)	—	2	—	—
Disulfoton	[298-04-4] (1986)	—	0.1	—	—
2,6-Di-tert-butyl-p-cresol	[128-37-0] (1987)	—	10	—	—
Diuron	[330-54-1] (1977)	—	10	—	—
Divinyl benzene	[1321-74-0] (1980)	10	53	—	—
Emery	[112-62-9] (1986)	—	10[(e)]	—	—
Endosulfan	[115-29-7] — Skin (1986)	—	0.1	—	—

Substance [CAS #]	ADOPTED VALUES TWA ppm[a)]	mg/m$^{3b)}$	STEL ppm[a)]	mg/m$^{3b)}$
Endrin [72-20-8] — Skin (1988)	—	0.1	—	—
Enflurane [13838-16-9] (1988)	75	566	—	—
Enzymes, *see* Subtilisins				
•■Epichlorohydrin [106-89-8] — Skin (1986) . . .	2	7.6	—	—
◂EPN [2104-64-5] — Skin (1986)	—	0.5	—	—
1,2-Epoxypropane, *see* Propylene oxide				
2,3-Epoxy-1-propanol, *see* Glycidol				
Ethane [74-84-0] (1981)	—(d)	—	—	—
Ethanethiol, *see* Ethyl mercaptan				
Ethanol, *see* Ethyl alcohol				
Ethanolamine [141-43-5] (1978)	3	7.5	6	15
◂Ethion [563-12-2] — Skin (1977)	—	0.4	—	—
•2-Ethoxyethanol [110-80-5] — Skin (1984) . . .	5	18	—	—
2-Ethoxyethyl acetate [111-15-9] — Skin (1984) . . .	5	27	—	—
Ethyl acetate [141-78-6] (1977)	400	1440	—	—
‡■Ethyl acrylate [140-88-5]	5	20	(25)	(100)
Ethyl alcohol [64-17-5] (1977) .	1000	1880	—	—
Ethylamine [75-04-7] (1977) . .	10	18	—	—
Ethyl amyl ketone [541-85-5] (1977)	25	131	—	—
◂Ethyl benzene [100-41-4] (1976)	100	434	125	543
Ethyl bromide [74-96-4] (1976)	200	891	250	1110
Ethyl butyl ketone [106-35-4] (1987)	50	234	—	—

(d) Simple asphyxiant; see definition in the "Introduction to the Chemical Substances."
(e) The value is for total dust containing no asbestos and < 1% crystalline silica.
Capital letters A & B refer to Appendices; C denotes ceiling limit.
( ) Adopted values enclosed are on the Notice of Intended Changes list. Consult it for currcnt proposal for change.
◂ Identifies substances for which there are also BEIs (*see* BEI section). Substances identified in the BEI documentations for methemoglobin inducers (for which methemoglobin is the principle toxicity) and organophosphorus cholinesterase inhibitors are part of this notation.
■ Substance identified by other sources as a suspected or confirmed human carcinogen. See the compliation in the Appendix to the Documentation of TLVs, pp. A-5(86)–A-9(86).
• Substance for which OSHA and/or NIOSH has a Permissible Exposure Limit (PEL) or a Recommended Exposure Limit (REL) lower than the TLV.
‡ See Notice of Intended Changes.

Substance [CAS #]	TWA ppm[a]	TWA mg/m^{3}[b]	STEL ppm[a]	STEL mg/m^{3}[b]
	ADOPTED VALUES			
Ethyl chloride [75-00-3] (1986)	1000	2640	—	—
Ethylene [74-85-1] (1981)	—[d]	—	—	—
Ethylene chlorohydrin [107-07-3]—Skin (1977) ...	C 1	C 3.3	—	—
Ethylenediamine [107-15-3] (1977)	10	25	—	—
▪Ethylene dibromide [106-93-4]—Skin (1982) ...	A2	A2	—	—
•▪Ethylene dichloride [107-06-2] (1986)	10	40	—	—
Ethylene glycol [107-21-1] Vapor (1981)	C 50	C 127	—	—
•Ethylene glycol dinitrate [628-96-6]—Skin (1985) ...	0.05	0.31	—	—
Ethylene glycol methyl ether acetate, *see* 2-Methoxyethyl acetate				
▪Ethylene oxide [75-21-8] (1984)	1,A2	1.8,A2	—	—
•▪Ethylenimine [151-56-4]—Skin (1977)	0.5	0.88	—	—
Ethyl ether [60-29-7] (1976) ..	400	1210	500	1520
Ethyl formate [109-94-4] (1987)	100	303	—	—
Ethylidene chloride, *see* 1,1-Dichloroethane				
Ethylidene norbornene [16219-75-3] (1977)	C 5	C 25	—	—
Ethyl mercaptan [75-08-1] (1986)	0.5	1.3	—	—
N-Ethylmorpholine [100-74-3]—Skin (1986) ...	5	24	—	—
Ethyl silicate [78-10-4] (1986) .	10	85	—	—
◂Fenamiphos [22224-92-6]—Skin (1984)	—	0.1	—	—
◂Fensulfothion [115-90-2] (1977)	—	0.1	—	—
◂Fenthion [55-38-9]—Skin (1983)	—	0.2	—	—
Ferbam [14484-64-1] (1986) ...	—	10	—	—
Ferrovanadium dust [12604-58-9] (1983)	—	1	—	3
•Fibrous glass dust (1978) ...	—	10	—	—
◂Fluorides, as F (1977)	—	2.5	—	—
•Fluorine [7782-41-4] (1976) ...	1	1.6	2	3.1
Fluorotrichloromethane, *see* Trichlorofluoromethane				
◂Fonofos [944-22-9]—Skin (1977)	—	0.1	—	—
‡•▪Formaldehyde [50-00-0]	(1,A2)	(1.2,A2)	(2,A2)	(2.5,A2)

Substance [CAS #]	ADOPTED VALUES			
	TWA		STEL	
	ppm[a)]	mg/m[3b)]	ppm[a)]	mg/m[3b)]
Formamide [75-12-7]— Skin (1988)	10	18	—	—
‡Formic acid [64-18-6]	5	9.4	(—)	(—)
◄Furfural [98-01-1]—Skin (1987)	2	7.9	—	—
Furfuryl alcohol [98-00-0]— Skin (1982)	10	40	15	60
•Gasoline [8006-61-9] (1982)	300	890	500	1480
Germanium tetrahydride [7782-65-2] (1986)	0.2	0.63	—	—
Glass, fibrous or dust, *see* Fibrous glass dust				
Glutaraldehyde [111-30-8] (1979)	C 0.2	C 0.82	—	—
Glycerin mist [56-81-5] (1981)	—	10[(e)]	—	—
Glycidol [556-52-5] (1987)	25	76	—	—
Glycol monoethyl ether, *see* 2-Ethoxyethanol				
Grain dust (oat, wheat, barley) (1986)	—	4[(i)]	—	—
‡Graphite (natural) [7782-42-5]	—	(2.5[(j)], Respirable dust)		
‡Graphite (synthetic)	—	(10 [(e)])	—	—
Gypsum, *see* Calcium sulfate				
Hafnium [7440-58-6] (1986)	—	0.5	—	—
Halothane [151-67-7] (1988)	50	404	—	—

(d) Simple asphyxiant; see definition in the "Introduction to the Chemical Substances."
(e) The value is for total dust containing no asbestos and < 1% crystalline silica.
(i) Total particulate.
(j) These TLVs are for the respirable fraction of dust for the substance listed. The concentration of respirable dust for the application of this limit is to be determined from the fraction passing a size-selector with the characteristics defined in the "c." paragraphs of Appendix D.

Capital letters A & B refer to Appendices; C denotes ceiling limit.

( ) Adopted values enclosed are on the Notice of Intended Changes list. Consult it for current proposal for change.

◄ Identifies substances for which there are also BEIs (*see* BEI section). Substances identified in the BEI documentations for methemoglobin inducers (for which methemoglobin is the principle toxicity) and organophosphorus cholinesterase inhibitors are part of this notation.

■ Substance identified by other sources as a suspected or confirmed human carcinogen. See the compliation in the Appendix to the Documentation of TLVs, pp. A-5(86)–A-9(86).

• Substance for which OSHA and/or NIOSH has a Permissible Exposure Limit (PEL) or a Recommended Exposure Limit (REL) lower than the TLV.

‡ See Notice of Intended Changes.

Substance [CAS #]	TWA ppm[a]	TWA mg/m³[b]	STEL ppm[a]	STEL mg/m³[b]
Helium [7440-59-7] (1981)	—[d]	—	—	—
▪Heptachlor [76-44-8]—Skin (1986)	—	0.5	—	—
Heptane [142-82-5] (n-Heptane) (1976)	400	1640	500	2050
2-Heptanone, *see* Methyl n-amyl ketone				
3-Heptanone, *see* Ethyl butyl ketone				
▪Hexachlorobutadiene [87-68-3]—Skin (1982)	0.02,A2	0.21,A2	—	—
Hexachlorocyclopentadiene [77-47-4] (1986)	0.01	0.11	—	—
*Hexachloroethane [67-72-1] (1989)	1	9.7	—	—
Hexachloronaphthalene [1335-87-1]—Skin (1986) ..	—	0.2	—	—
Hexafluoroacetone [684-16-2]—Skin (1986) ...	0.1	0.68	—	—
Hexamethylene diisocyanate [822-06-0] (1988)	0.005	0.034	—	—
▪Hexamethyl phosphoramide [680-31-9]—Skin (1978) ...	A2	A2	—	—
◂Hexane (n-Hexane) [110-54-3] (1982)	50	176	—	—
• Other isomers (1982).....	500	1760	1000	3500
2-Hexanone, *see* Methyl n-butyl ketone				
Hexone, *see* Methyl isobutyl ketone				
sec-Hexyl acetate [108-84-9] (1977)	50	295	—	—
Hexylene glycol [107-41-5] (1977)	C 25	C 121	—	—
‡•▪Hydrazine [302-01-2]—Skin..	(0.1,A2)	(0.13,A2)	—	—
Hydrogen [1333-74-0] (1981) ..	—[d]	—	—	—
Hydrogenated terphenyls [61788-32-7] (1977)	0.5	4.9	—	—
Hydrogen bromide [10035-10-6] (1986)	C 3	C 9.9	—	—
Hydrogen chloride [7647-01-0] (1977)	C 5	C 7.5	—	—
•Hydrogen cyanide [74-90-8]—Skin (1980)	C 10	C 11	—	—
Hydrogen fluoride [7664-39-3], as F (1986) ...	C 3	C 2.6	—	—

Substance [CAS #]	ADOPTED VALUES TWA ppm[a)]	mg/m[3b)]	STEL ppm[a)]	mg/m[3b)]
Hydrogen peroxide [7722-84-1] (1986)	1	1.4	—	—
Hydrogen selenide [7783-07-5], as Se (1977) ..	0.05	0.16	—	—
•Hydrogen sulfide [7783-06-4] (1976)	10	14	15	21
Hydroquinone [123-31-9] (1987)	—	2	—	—
4-Hydroxy-4-methyl-2-pentanone, see Diacetone alcohol				
2-Hydroxypropyl acrylate [999-61-1] — Skin (1980) ..	0.5	2.8	—	—
Indene [95-13-6] (1987)	10	48	—	—
Indium [7440-74-6] & compounds, as In (1986).	—	0.1	—	—
Iodine [7553-56-2] (1977)	C 0.1	C 1.0	—	—
Iodoform [75-47-8] (1986) ...	0.6	10	—	—
Iron oxide fume (Fe_2O_3) [1309-37-1], as Fe (1986) ..	B2	5	—	—
Iron pentacarbonyl [13463-40-6], as Fe (1982) .	0.1	0.23	0.2	0.45
Iron salts, soluble, as Fe (1986)	—	1	—	—
Isoamyl acetate [123-92-2] (1987)	100	532	—	—
Isoamyl alcohol [123-51-3] (1976)	100	361	125	452
‡Isobutyl acetate [110-19-0] ..	150	713	(187)	(888)

(d) Simple asphyxiant; see definition in the "Introduction to the Chemical Substances." Capital letters A & B refer to Appendices; C denotes ceiling limit.

( ) Adopted values enclosed are on the Notice of Intended Changes list. Consult it for current proposal for change.

◄ Identifies substances for which there are also BEIs (*see* BEI section). Substances identified in the BEI documentations for methemoglobin inducers (for which methemoglobin is the principle toxicity) and organophosphorus cholinesterase inhibitors are part of this notation.

■ Substance identified by other sources as a suspected or confirmed human carcinogen. See the compliation in the Appendix to the Documentation of TLVs, pp. A-5(86)–A-9(86).

• Substance for which OSHA and/or NIOSH has a Permissible Exposure Limit (PEL) or a Recommended Exposure Limit (REL) lower than the TLV.

‡ See Notice of Intended Changes.

* 1989–1990 Adoption.

Substance [CAS #]	ADOPTED VALUES TWA ppm[a)]	TWA mg/m³[b)]	STEL ppm[a)]	STEL mg/m³[b)]
Isobutyl alcohol [78-83-1] (1987)	50	152	—	—
Isooctyl alcohol [26952-21-6]—Skin (1982)	50	266	—	—
Isophorone [78-59-1] (1977)	C 5	C 28	—	—
Isophorone diisocyanate [4098-71-9] — Skin (1988)	0.005	0.045	—	—
Isopropoxyethanol [109-59-1] (1987)	25	106	—	—
Isopropyl acetate [108-21-4] (1976)	250	1040	310	1290
Isopropyl alcohol [67-63-0] (1976)	400	985	500	1230
Isopropylamine [75-31-0] (1976)	5	12	10	24
N-Isopropylaniline [768-52-5] — Skin (1986)	2	11	—	—
Isopropyl ether [108-20-3] (1976)	250	1040	310	1300
•Isopropyl glycidyl ether (IGE) [4016-14-2] (1976)	50	238	75	356
Kaolin (1986)	—	10[(e)]	—	—
Ketene [463-51-4] (1976)	0.5	0.86	1.5	2.6
▪•◂Lead [7439-92-1], inorg. dusts & fumes, as Pb (1986)	—	0.15	—	—
Lead arsenate [3687-31-8], as $Pb_3(AsO_4)_2$ (1985)	—	0.15	—	—
‡•▪Lead chromate [7758-97-6], (as Cr)	—	0.05,A2	—	—
Limestone, *see* Calcium carbonate				
▪Lindane [58-89-9] — Skin (1986)	—	0.5	—	—
Lithium hydride [7580-67-8] (1977)	—	0.025	—	—
L.P.G. (Liquified petroleum gas) [68476-85-7] (1987)	1000	1800	—	—
Magnesite [546-93-0] (1986)	—	10[(e)]	—	—
Magnesium oxide fume [1309-48-4] (1977)	—	10	—	—
◂Malathion [121-75-5] — Skin (1977)	—	10	—	—
Maleic anhydride [108-31-6] (1977)	0.25	1.0	—	—

Substance [CAS #]	TWA ppm[a)]	TWA mg/m^{3}[b)]	STEL ppm[a)]	STEL mg/m^{3}[b)]
Manganese [7439-96-5], as Mn				
• Dust & compounds (1988)	—	5	—	—
Fume (1979)	—	1	—	3
Manganese cyclopentadienyl tricarbonyl [12079-65-1], as Mn—Skin (1986)	—	0.1	—	—
Marble, *see* Calcium carbonate				
Mercury [7439-97-6], as Hg—Skin				
Alkyl compounds (1980) .	—	0.01	—	0.03
All forms except alkyl				
Vapor (1982)	—	0.05	—	—
• Aryl & inorganic compounds (1982) ..	—	0.1	—	—
•Mesityl oxide [141-79-7] (1981)	15	60	25	100
Methacrylic acid [79-41-4] (1981)	20	70	—	—
Methane [74-82-8] (1981)....	—(d)	—	—	—
Methanethiol, *see* Methyl mercaptan				
Methanol, *see* Methyl alcohol				
◂Methomyl [16752-77-5] (1977)	—	2.5	—	—
Methoxychlor [72-43-5] (1977)	—	10	—	—
•2-Methoxyethanol [109-86-4]—Skin (1984)...	5	16	—	—
2-Methoxyethyl acetate [110-49-6]—Skin (1984)...	5	24	—	—
4-Methoxyphenol [150-76-5] (1982).........	—	5	—	—

(d) Simple asphyxiant; see definition in the "Introduction to the Chemical Substances."
(e) The value is for total dust containing no asbestos and < 1% crystalline silica.
Capital letters A & B refer to Appendices; C denotes ceiling limit.
( ) Adopted values enclosed are on the Notice of Intended Changes list. Consult it for current proposal for change.
◂ Identifies substances for which there are also BEIs (*see* BEI section). Substances identified in the BEI documentations for methemoglobin inducers (for which methemoglobin is the principle toxicity) and organophosphorus cholinesterase inhibitors are part of this notation.
■ Substance identified by other sources as a suspected or confirmed human carcinogen. See the compliation in the Appendix to the Documentation of TLVs, pp. A-5(86)–A-9(86).
• Substance for which OSHA and/or NIOSH has a Permissible Exposure Limit (PEL) or a Recommended Exposure Limit (REL) lower than the TLV.
‡ See Notice of Intended Changes.

		ADOPTED VALUES			
		TWA		STEL	
Substance	**[CAS #]**	**ppm[a)]**	**mg/m³[b)]**	**ppm[a)]**	**mg/m³[b)]**
Methyl acetate [79-20-9] (1976)		200	606	250	760
‡Methyl acetylene [74-99-7] . .		1000	1640	(1250)	(2050)
Methyl acetylene-propadiene mixture (MAPP) (1976) . .		1000	1640	1250	2050
Methyl acrylate [96-33-3] — Skin (1977)		10	35	—	—
Methylacrylonitrile [126-98-7] — Skin (1986) . . .		1	2.7	—	—
Methylal [109-87-5] (1987) . . .		1000	3110	—	—
Methyl alcohol [67-56-1] — Skin (1976)		200	262	250	328
Methylamine [74-89-5] (1977)		10	13	—	—
Methyl amyl alcohol, *see* Methyl isobutyl carbinol					
Methyl n-amyl ketone [110-43-0] (1987)		50	233	—	—
◂N-Methyl aniline [100-61-8]— Skin (1986)		0.5	2.2	—	—
•▪Methyl bromide [74-83-9] — Skin (1986)		5	19	—	—
•Methyl n-butyl ketone [591-78-6] — Skin (1981) . . .		5	20	—	—
•▪Methyl chloride [74-87-3] — Skin (1981)		50	103	100	207
•◂Methyl chloroform [71-55-6] (1976)		350	1910	450	2460
Methyl 2-cyanoacrylate [137-05-3] (1976)		2	9.1	4	18
Methylcyclohexane [108-87-2] (1987)		400	1610	—	—
Methylcyclohexanol [25639-42-3] (1987)		50	234	—	—
o-Methylcyclohexanone [583-60-8] — Skin (1976) . . .		50	229	75	344
2-Methylcyclopentadienyl manganese tricarbonyl [12108-13-3], as Mn — Skin (1986)		—	0.2	—	—
◂Methyl demeton [8022-00-2] — Skin (1986) . .		—	0.5	—	—
Methylene bisphenyl isocyanate (MDI) [101-68-8] (1988)		0.005	0.051	—	—

Substance	[CAS #]	ADOPTED VALUES TWA ppm[a)]	mg/m$^{3b)}$	STEL ppm[a)]	mg/m$^{3b)}$
■•Methylene chloride	[75-09-2] (1988)	50,A2	174,A2	—	—
•■◄4,4′-Methylene bis(2-chloroaniline)	[101-14-4]—Skin (1974)	0.02,A2	0.22,A2	—	—
Methylene bis(4-cyclohexylisocyanate)	[5124-30-1] (1988)	0.005	0.054	—	—
■4,4′-Methylene dianiline	[101-77-9]—Skin (1986)	0.1,A2	0.81,A2	—	—
◄Methyl ethyl ketone (MEK)	[78-93-3] (1976)	200	590	300	885
Methyl ethyl ketone peroxide	[1338-23-4] (1977)	C 0.2	C 1.5	—	—
Methyl formate	[107-31-3] (1976)	100	246	150	369
5-Methyl-3-heptanone, *see* Ethyl amyl ketone					
‡•■Methyl hydrazine	[60-34-4]—Skin	(C 0.2,A2)	(C 0.38,A2)	—	—
•■Methyl iodide	[74-88-4]—Skin (1986)	2,A2	12,A2	—	—
Methyl isoamyl ketone	[110-12-3] (1982)	50	234	—	—
Methyl isobutyl carbinol	[108-11-2]—Skin (1976)	25	104	40	166
Methyl isobutyl ketone	[108-10-1] (1981)	50	205	75	308
Methyl isocyanate	[624-83-9]—Skin (1977)	0.02	0.047	—	—

Capital letters A & B refer to Appendices; C denotes ceiling limit.

( ) Adopted values enclosed are on the Notice of Intended Changes list. Consult it for current proposal for change.

◄ Identifies substances for which there are also BEIs (*see* BEI section). Substances identified in the BEI documentations for methemoglobin inducers (for which methemoglobin is the principle toxicity) and organophosphorus cholinesterase inhibitors are part of this notation.

■ Substance identified by other sources as a suspected or confirmed human carcinogen. See the compliation in the Appendix to the Documentation of TLVs, pp. A-5(86)–A-9(86).

• Substance for which OSHA and/or NIOSH has a Permissible Exposure Limit (PEL) or a Recommended Exposure Limit (REL) lower than the TLV.

‡ See Notice of Intended Changes.

	ADOPTED VALUES			
	TWA		STEL	
Substance [**CAS #**]	ppm[a)]	mg/m^{3}[b)]	ppm[a)]	mg/m^{3}[b)]
Methyl isopropyl ketone [563-80-4] (1981)	200	705	—	—
•Methyl mercaptan [74-93-1] (1977)	0.5	0.98	—	—
Methyl methacrylate [80-62-6] (1987)	100	410	—	—
◂Methyl parathion [298-00-0] — Skin (1986) ...	—	0.2	—	—
•Methyl propyl ketone [107-87-9] (1976)	200	705	250	881
Methyl silicate [681-84-5] (1986)	1	6.0	—	—
α-Methyl styrene [98-83-9] (1981)	50	242	100	484
Metribuzin [21087-64-9] (1984)	—	5	—	—
◂Mevinphos [7786-34-7] — Skin (1976)	0.01	0.09	0.03	0.27
Mica [12001-26-2] (1986)	—	3,(j) Respirable dust		
Mineral wool fiber (1974) ..	—	10(e)	—	—
Molybdenum [7439-98-7], as Mo				
Soluble compounds (1986)	—	5	—	—
Insoluble compounds (1986)	—	10	—	—
Monochlorobenzene, *see* Chlorobenzene				
Monocrotophos [6923-22-4] (1977)	—	0.25	—	—
‡Morpholine [110-91-8] — Skin	20	71	(30)	(107)
◂Naled [300-76-5] — Skin (1986)	—	3	—	—
Naphthalene [91-20-3] (1976).	10	52	15	79
▪β-Naphthylamine [91-59-8] (1972)	—	A1	—	—
Neon [7440-01-9] (1981)	—(d)	—	—	—
▪Nickel [7440-02-0]				
‡Metal	—	(1)	—	—
‡Insoluble compounds, as Ni	—	(1)	—	—
‡▪•Soluble compounds, as Ni	—	(0.1)	—	—
‡•▪Nickel carbonyl [13463-39-3], as Ni	(0.05)	(0.12)	—	—
‡•▪Nickel sulfide roasting, fume & dust, as Ni	—	(1,A1)	—	—

Substance [CAS #]	ADOPTED VALUES			
	TWA		STEL	
	ppm[a)]	mg/m$^{3b)}$	ppm[a)]	mg/m$^{3b)}$
Nicotine [54-11-5] — Skin (1986)	—	0.5	—	—
Nitrapyrin [1929-82-4] (1982)	—	10	—	20
Nitric acid [7697-37-2] (1976)	2	5.2	4	10
◄Nitric oxide [10102-43-9] (1986)	25	31	—	—
◄p-Nitroaniline [100-01-6] — Skin (1982)	—	3	—	—
◄Nitrobenzene [98-95-3] — Skin (1986)	1	5.0	—	—
◄p-Nitrochlorobenzene [100-00-5]—Skin (1988)	0.1	0.64	—	—
■4-Nitrodiphenyl [92-93-3] (1976)	—	A1	—	—
Nitroethane [79-24-3] (1986)	100	307	—	—
Nitrogen [7727-37-9] (1989)	—(d)	—	—	—
•Nitrogen dioxide [10102-44-0] (1981)	3	5.6	5	9.4
◄Nitrogen trifluoride [7783-54-2] (1986)	10	29	—	—
•Nitroglycerin (NG) [55-63-0] — Skin (1985)	0.05	0.46	—	—
Nitromethane [75-52-5] (1986)	100	250	—	—

(d) Simple asphyxiant; see definition in the "Introduction to the Chemical Substances."

(e) The value is for total dust containing no asbestos and < 1% crystalline silica.

(j) These TLVs are for the respirable fraction of dust for the substance listed. The concentration of respirable dust for the application of this limit is to be determined from the fraction passing a size-selector with the characteristics defined in the "c." paragraphs of Appendix D.

Capital letters A & B refer to Appendices; C denotes ceiling limit.

( ) Adopted values enclosed are on the Notice of Intended Changes list. Consult it for current proposal for change.

◄ Identifies substances for which there are also BEIs (*see* BEI section). Substances identified in the BEI documentations for methemoglobin inducers (for which methemoglobin is the principle toxicity) and organophosphorus cholinesterase inhibitors are part of this notation.

■ Substance identified by other sources as a suspected or confirmed human carcinogen. See the compliation in the Appendix to the Documentation of TLVs, pp. A-5(86)–A-9(86).

• Substance for which OSHA and/or NIOSH has a Permissible Exposure Limit (PEL) or a Recommended Exposure Limit (REL) lower than the TLV.

‡ See Notice of Intended Changes.

Substance [CAS #]	ADOPTED VALUES TWA ppm[a)]	mg/m³[b)]	STEL ppm[a)]	mg/m³[b)]
1-Nitropropane [108-03-2] (1986)	25	91	—	—
▪2-Nitropropane [79-46-9] (1987)	10,A2	36,A2	—	—
▪N-Nitrosodimethylamine [62-75-9] — Skin (1972)	—	A2	—	—
◂Nitrotoluene [88-72-2; 99-08-1; 99-99-0] — Skin (1982)	2	11	—	—
Nitrotrichloromethane, *see* Chloropicrin				
*Nitrous oxide [10028-97-2] (1989)	50	90	—	—
Nonane [111-84-2]	200	1050	—	—
Nuisance particulates, *see* Particulates Not Otherwise Classified (PNOC)				
Octachloronaphthalene [2234-13-1] — Skin (1976)	—	0.1	—	0.3
Octane [111-65-9] (1976)	300	1400	375	1750
Oil mist, mineral (1976)	—	5[(k)]	—	10
Osmium tetroxide [20816-12-0], as Os (1976)	0.0002	0.0016	0.0006	0.0048
Oxalic acid [144-62-7] (1976)	—	1	—	2
Oxygen difluoride [7783-41-7] (1986)	C 0.05	C 0.11	—	—
*Ozone [10028-15-6] (1989)	C 0.1	C 0.20	—	—
Paraffin wax fume [8002-74-2] (1987)	—	2	—	—
Paraquat [4685-14-7], respirable sizes (1978)	—	0.1	—	—
•◂Parathion [56-38-2] — Skin (1986)	—	0.1	—	—
Particulate polycyclic aromatic hydrocarbons (PPAH), *see* Coal tar pitch volatiles				
Particulates Not Otherwise Classified (PNOC) (1986)	—	10[(e)]	—	—
Pentaborane [19624-22-7] (1976)	0.005	0.01	0.013	0.03
Pentachloronaphthalene [1321-64-8] (1986)	—	0.5	—	—
◂Pentachlorophenol [87-86-5] — Skin (1986)	—	0.5	—	—
Pentaerythritol [115-77-5] (1986)	—	10[(e)]	—	—
•Pentane [109-66-0] (1976)	600	1770	750	2210

		ADOPTED VALUES			
		TWA		STEL	
Substance	[CAS #]	ppm[a)]	mg/m³[b)]	ppm[a)]	mg/m³[b)]
2-Pentanone, *see* Methyl propyl ketone					
■•◄Perchloroethylene	[127-18-4] (1984)	50	339	200	1368
Perchloromethyl mercaptan	[594-42-3] (1977)	0.1	0.76	—	—
Perchloryl fluoride	[7616-94-6] (1976)	3	13	6	25
Precipitated silica, *see* Silica—Amorphous					
Perlite	(1986)	—	10[(e)]	—	—
Petroleum distillates, *see* Gasoline; Stoddard solvent; VM&P naphtha					
Phenacyl chloride, *see* α-Chloroacetophenone					
◄Phenol [108-95-2]—Skin	(1987)	5	19	—	—
Phenothiazine [92-84-2]—Skin	(1986)	—	5	—	—
■N-Phenyl-beta-naphthylamine	[135-88-6] (1979)	A2	A2	—	—
‡p-Phenylenediamine	[106-50-3](—Skin)	—	0.1	—	—
Phenyl ether [101-84-8], vapor	(1976)	1	7.0	2	14
Phenylethylene, *see* Styrene, monomer					
•■Phenyl glycidyl ether (PGE)	[122-60-1] (1982)	1	6.1	—	—
‡•■Phenylhydrazine	[100-63-0]—Skin	(5,A2)	(22,A2)	(10,A2)	(44,A2)

(e) The value is for total dust containing no asbestos and < 1% crystalline silica.
(k) As sampled by method that does not collect vapor.
Capital letters A & B refer to Appendices; C denotes ceiling limit.
( ) Adopted values enclosed are on the Notice of Intended Changes list. Consult it for current proposal for change.
◄ Identifies substances for which there are also BEIs (*see* BEI section). Substances identified in the BEI documentations for methemoglobin inducers (for which methemoglobin is the principle toxicity) and organophosphorus cholinesterase inhibitors are part of this notation.
■ Substance identified by other sources as a suspected or confirmed human carcinogen. See the compliation in the Appendix to the Documentation of TLVs, pp. A-5(86)–A-9(86).
• Substance for which OSHA and/or NIOSH has a Permissible Exposure Limit (PEL) or a Recommended Exposure Limit (REL) lower than the TLV.
‡ See Notice of Intended Changes.
* 1989–1990 Adoption.

Substance [CAS #]	TWA ppm[a]	TWA mg/m^{3}[b]	STEL ppm[a]	STEL mg/m^{3}[b]
	ADOPTED VALUES			
•Phenyl mercaptan [108-98-5] (1978)	0.5	2.3	—	—
Phenylphosphine [638-21-1] (1977)	C 0.05	C 0.23	—	—
Phorate [298-02-2] — Skin (1976)	—	0.05	—	0.2
Phosdrin, *see* Mevinphos				
Phosgene [75-44-5] (1978)	0.1	0.40	—	—
Phosphine [7803-51-2] (1976)	0.3	0.42	1	1.4
Phosphoric acid [7664-38-2] (1976)	—	1	—	3
Phosphorus (yellow) [7723-14-0] (1986)	—	0.1	—	—
‡Phosphorus oxychloride [10025-87-3]	0.1	0.63	(0.5)	(3.1)
Phosphorus pentachloride [10026-13-8] (1980)	0.1	0.85	—	—
Phosphorus pentasulfide [1314-80-3] (1976)	—	1	—	3
Phosphorus trichloride [7719-12-2] (1982)	0.2	1.1	0.5	2.8
Phthalic anhydride [85-44-9] (1987)	1	6.1	—	—
m-Phthalodinitrile [626-17-5] (1977)	—	5	—	—
‡Picloram [1918-02-1]	—	10	—	(20)
‡Picric acid [88-89-1] — Skin	—	0.1	—	(0.3)
Pindone [83-26-1] (1987)	—	0.1	—	—
Piperazine dihydrochloride [142-64-3] (1982)	—	5	—	—
2-Pivalyl-1,3-indandione, *see* Pindone				
Plaster of Paris, *see* Calcium sulfate				
Platinum [7440-06-4]				
Metal (1981)	—	1	—	—
Soluble salts, as Pt (1981)	—	0.002	—	—
Polychlorobiphenyls, *see* Chlorodiphenyls				
Polytetrafluoroethylene decomposition products (1972)	—	B1	—	—
Portland cement (1986)	—	10[e]	—	—
Potassium hydroxide [1310-58-3] (1977)	—	C 2	—	—

		ADOPTED VALUES			
		TWA		STEL	
Substance	[CAS #]	ppm[a)]	mg/m[3b)]	ppm[a)]	mg/m[3b)]
•Propane [74-98-6] (1981)		—(d)	—	—	—
▪Propane sultone [1120-71-4] (1977)		A2	A2	—	—
Propargyl alcohol [107-19-7] — Skin (1987)		1	2.3	—	—
▪β-Propiolactone [57-57-8] (1987)		0.5,A2	1.5,A2	—	—
‡Propionic acid [79-09-4]		10	30	(15)	(45)
Propoxur [114-26-1] (1987)		—	0.5	—	—
n-Propyl acetate [109-60-4] (1976)		200	835	250	1040
n-Propyl alcohol [71-23-8] — Skin (1976)		200	492	250	615
Propylene [115-07-1] (1976)		—(d)	—	—	—
Propylene dichloride [78-87-5] (1976)		75	347	110	509
◂Propylene glycol dinitrate [6423-43-4] — Skin (1985)		0.05	0.34	—	—
Propylene glycol monomethyl ether [107-98-2] (1976)		100	369	150	553
▪Propylene imine [75-55-8] — Skin (1983)		2,A2	4.7,A2	—	—
▪Propylene oxide [75-56-9] (1981)		20	48	—	—
◂n-Propyl nitrate [627-13-4] (1978)		25	107	40	172

(d) Simple asphyxiant; see definition in the "Introduction to the Chemical Substances."
(e) The value is for total dust containing no asbestos and < 1% crystalline silica.
Capital letters A & B refer to Appendices; C denotes ceiling limit.
( ) Adopted values enclosed are on the Notice of Intended Changes list. Consult it for current proposal for change.
◂ Identifies substances for which there are also BEIs (*see* BEI section). Substances identified in the BEI documentations for methemoglobin inducers (for which methemoglobin is the principle toxicity) and organophosphorus cholinesterase inhibitors are part of this notation.
▪ Substance identified by other sources as a suspected or confirmed human carcinogen. See the compliation in the Appendix to the Documentation of TLVs, pp. A-5(86)–A-9(86).
• Substance for which OSHA and/or NIOSH has a Permissible Exposure Limit (PEL) or a Recommended Exposure Limit (REL) lower than the TLV.
‡ See Notice of Intended Changes.

Substance [CAS #]	ADOPTED VALUES TWA ppm[a)]	mg/m^3[b)]	STEL ppm[a)]	mg/m^3[b)]
Propyne, *see* Methyl acetylene				
Pyrethrum [8003-34-7] (1981)	—	5	—	—
Pyridine [110-86-1] (1987) . . .	5	16	—	—
Pyrocatechol, *see* Catechol				
Quartz, *see* Silica—Crystaline				
Quinone [106-51-4] (1987) . . .	0.1	0.44	—	—
RDX, *see* Cyclonite				
Resorcinol [108-46-3] (1976) .	10	45	20	90
•Rhodium [7440-16-6]				
•Metal (1982)	—	1	—	—
•Insoluble compounds, as Rh (1984)	—	1	—	—
•Soluble compounds, as Rh (1984)	—	0.01	—	—
Ronnel [299-84-3] (1977)	—	10	—	—
Rosin core solder pyrolysis products, as formaldehyde (1987)	—	0.1	—	—
Rotenone (commercial) [83-79-4] (1987)	—	5	—	—
Rouge (1986)	—	10[(e)]	—	—
•Rubber solvent (Naphtha) (1977)	400	1570	—	—
Selenium compounds [7782-49-2], as Se (1977) . .	—	0.2	—	—
Selenium hexafluoride [7783-79-1], as Se (1979) . .	0.05	0.16	—	—
Sesone [136-78-7] (1986)	—	10	—	—
Silane, *see* Silicon tetrahydride				
Silica—Amorphous				
•Diatomaceous earth (uncalcined) [61790-53-2] (1986)	—	10[(e)]	—	—
•Precipitated silica (1987)	—	10[(e)]	—	—
•Silica gel (1987)	—	10[(e)]	—	—
▪Silica—Crystalline				
Cristobalite [14464-46-1] (1986)	—	0.05,[(j)] Respirable dust		
•Quartz [14808-60-7] (1986)	—	0.1,[(j)] Respirable dust		
•Silica, fused [7631-86-9] (1985)	—	0.1,[(j)] Respirable dust		

		ADOPTED VALUES			
		TWA		STEL	
Substance	[CAS #]	ppm[a)]	mg/m[3b)]	ppm[a)]	mg/m[3b)]
Tridymite [15468-32-3] (1986)		—	0.05,[(j)]	Respirable dust	
•Tripoli [1317-95-9] (1985)		—	0.1,[(j)]	of contained respirable quartz	
Silica, fused, *see* Silica—Crystaline					
Silica, gel, *see* Silica—Amorphous					
Silicon [7440-21-3] (1986)		—	10[(e)]	—	—
Silicon carbide [409-21-2] (1986)		—	10[(e)]	—	—
Silicon tetrahydride [7803-62-5] (1983)		5	6.6	—	—
Silver [7440-22-4]					
•Metal (1981)		—	0.1	—	—
Soluble compounds, as Ag (1981)		—	0.01	—	—
Soapstone					
Respirable dust (1985)		—	3[(j)]	—	—
Total dust (1985)		—	6[(e)]	—	—
Sodium azide [26628-22-8] (1977)		C 0.11	C 0.3	—	—
Sodium bisulfite [7631-90-5] (1980)		—	5	—	—
Sodium 2,4-dichloro-phenoxyethyl sulfate, *see* Sesone					
Sodium fluoroacetate [62-74-8] — Skin (1976)		—	0.05	—	0.15
Sodium hydroxide [1310-73-2] (1977)		—	C 2	—	—
Sodium metabisulfite [7681-57-4] (1980)		—	5	—	—

(e) The value is for total dust containing no asbestos and < 1% crystalline silica.
(j) These TLVs are for the respirable fraction of dust for the substance listed. The concentration of respirable dust for the application of this limit is to be determined from the fraction passing a size-selector with the characteristics defined in the "c." paragraphs of Appendix D.

Capital letters A & B refer to Appendices; C denotes ceiling limit.

■ Substance identified by other sources as a suspected or confirmed human carcinogen. See the compliation in the Appendix to the Documentation of TLVs, pp. A-5(86)–A-9(86).

• Substance for which OSHA and/or NIOSH has a Permissible Exposure Limit (PEL) or a Recommended Exposure Limit (REL) lower than the TLV.

Substance [CAS #]	TWA ppm[a)]	TWA mg/m³[b)]	STEL ppm[a)]	STEL mg/m³[b)]
	ADOPTED VALUES			
Starch [9005-25-8] (1986)	—	10[(e)]	—	—
Stearates[(l)] (1988)	—	10[(e)]	—	—
Stibine [7803-52-3] (1986)	0.1	0.51	—	—
•Stoddard solvent [8052-41-3] (1987)	100	525	—	—
Strychnine [57-24-9] (1986)	—	0.15	—	—
▪◂Styrene, monomer [100-42-5] — Skin (1981)	50	213	100	426
Subtilisins [1395-21-7] (Proteolytic enzymes as 100% pure crystalline enzyme) (1977)	—	C0.00006[(m)]	—	—
Sucrose [57-50-1] (1986)	—	10[(e)]	—	—
◂Sulfotep [3689-24-5] — Skin (1986)	—	0.2	—	—
•Sulfur dioxide [7446-09-5] (1986)	2	5.2	5	13
Sulfur hexafluoride [2551-62-4] (1986)	1000	5970	—	—
*Sulfuric acid [7664-93-9] (1989)	—	1	—	3
Sulfur monochloride [10025-67-9] (1986)	C 1	C 5.5	—	—
Sulfur pentafluoride [5714-22-7] (1986)	C 0.01	C 0.10	—	—
Sulfur tetrafluoride [7783-60-0] (1986)	C 0.1	C 0.44	—	—
Sulfuryl fluoride [2699-79-8] (1976)	5	21	10	42
Sulprofos [35400-43-2] (1984)	—	1	—	—
Systox, *see* Demeton				
2,4,5-T [93-76-5] (1986)	—	10	—	—
▪Talc (containing no abestos fibers) [14807-96-6] (1983)	—	2,[(j)] Respirable dust		
Talc (containing asbestos fibers) (1985)	Use asbestos TLV-TWA.[(n)]			
Tantalum [7440-25-7], metal & oxide dusts (1988)	—	5	—	—
TEDP, *see* Sulfotep				
Tellurium & compounds [13494-80-9], as Te (1977)	—	0.1	—	—

		ADOPTED VALUES			
		TWA		STEL	
Substance	**[CAS #]**	**ppm**[a)]	**mg/m**[3b)]	**ppm**[a)]	**mg/m**[3b)]
Tellurium hexafluoride [7783-80-4], as Te (1977) . .		0.02	0.10	—	—
◂Temephos [3383-96-8] (1986)		—	10	—	—
◂TEPP [107-49-3] — Skin (1986)		0.004	0.047	—	—
Terphenyls [26140-60-3] (1980)		C 0.5	C 4.7	—	—
1,1,1,2-Tetrachloro-2,2-difluoroethane [76-11-9] (1986)		500	4170	—	—
1,1,2,2-Tetrachloro-1,2-difluoroethane [76-12-0] (1986)		500	4170	—	—
▪1,1,2,2-Tetrachloroethane [79-34-5] — Skin (1986) . . .		1	6.9	—	—
Tetrachloroethylene, *see* Perchloroethylene					
Tetrachloromethane, *see* Carbon tetrachloride					
Tetrachloronaphthalene [1335-88-2] (1986)		—	2	—	—
•Tetraethyl lead [78-00-2], as Pb — Skin (1986)		—	0.1[(o)]	—	—
Tetrahydrofuran [109-99-9] (1976)		200	590	250	738

(e) The value is for total dust containing no asbestos and < 1% crystalline silica.

(j) These TLVs are for the respirable fraction of dust for the substance listed. The concentration of respirable dust for the application of this limit is to be determined from the fraction passing a size-selector with the characteristics defined in the "c." paragraphs of Appendix D.

(m) Based on "high volume" sampling.

(n) However, should not exceed 2 mg/m³ respirable dust.

(o) For control of general room air, biologic monitoring is essential for personnel control.

Capital letters A & B refer to Appendices; C denotes ceiling limit.

◂ Identifies substances for which there are also BEIs (*see* BEI section). Substances identified in the BEI documentations for methemoglobin inducers (for which methemoglobin is the principle toxicity) and organophosphorus cholinesterase inhibitors are part of this notation.

▪ Substance identified by other sources as a suspected or confirmed human carcinogen. See the compliation in the Appendix to the Documentation of TLVs, pp. A-5(86)–A-9(86).

• Substance for which OSHA and/or NIOSH has a Permissible Exposure Limit (PEL) or a Recommended Exposure Limit (REL) lower than the TLV.

* 1989–1990 Adoption.

Substance [CAS #]	TWA ppm[a)]	TWA mg/m³[b)]	STEL ppm[a)]	STEL mg/m³[b)]
	ADOPTED VALUES			
•Tetramethyl lead [75-74-1], as Pb — Skin (1986)	—	0.15[(o)]	—	—
Tetramethyl succinonitrile [3333-52-6] — Skin (1986)	0.5	2.8	—	—
Tetranitromethane [509-14-8] (1977)	1	8.0	—	—
Tetrasodium pyrophosphate [7722-88-5] (1980)	—	5	—	—
•Tetryl [479-45-8] (1986)	—	1.5	—	—
Thallium [7440-28-0]				
Soluble compounds, as Tl — Skin (1977)	—	0.1	—	—
4,4′-Thiobis(6-tert-butyl-m-cresol) [96-69-5] (1986)	—	10	—	—
Thioglycolic acid [68-11-1] — Skin (1978)	1	3.8	—	—
Thionyl chloride [7719-09-7] (1986)	C 1	C 4.9	—	—
‡Thiram [137-26-8]	—	(5)	—	—
Tin [7440-31-5]				
Metal (1986)	—	2	—	—
Oxide & inorganic compounds, except SnH_4, as Sn (1986)	—	2	—	—
‡Organic compounds, as Sn — Skin	—	0.1	—	(—)
Titanium dioxide [13463-67-7] (1986)	—	10[(e)]	—	—
▪o-Tolidine [119-93-7] — Skin (1982)	A2	A2	—	—
◂Toluene [108-88-3] (1976)	100	377	150	565
▪Toluene-2,4-diisocyanate (TDI) [584-84-9] (1983)	0.005	0.036	0.02	0.14
▪◂o-Toluidine [95-53-4] — Skin (1984)	2,A2	8.8,A2	—	—
◂m-Toluidine [108-44-1] — Skin (1986)	2	8.8	—	—
◂p-Toluidine [106-49-0] — Skin (1986)	2,A2	8.8,A2	—	—
Toluol, *see* Toluene				
Toxaphene, *see* Chlorinated camphene				

Substance	[CAS #]	TWA ppm[a)]	TWA mg/m^3[b)]	STEL ppm[a)]	STEL mg/m^3[b)]
		ADOPTED VALUES			
Tributyl phosphate	[126-73-8] (1986)	0.2	2.2	—	—
Trichloroacetic acid	[76-03-9] (1980)	1	6.7	—	—
1,2,4-Trichlorobenzene	[120-82-1] (1978)	C 5	C 37	—	—
1,1,1-Trichloroethane, *see* Methyl chloroform					
•■1,1,2-Trichloroethane	[79-00-5]—Skin (1986)	10	55	—	—
•■◄Trichloroethylene	[79-01-6] (1984)	50	269	200	1070
Trichlorofluoromethane	[75-69-4] (1982)	C 1000	C 5620	—	—
Trichloromethane, *see* Chloroform					
Trichloronaphthalene	[1321-65-9]—Skin (1986)	—	5	—	—
Trichloronitromethane, *see* Chloropicrin					
1,2,3-Trichloropropane	[96-18-4]—Skin (1987)	10	60	—	—
1,1,2-Trichloro-1,2,2-trifluoroethane	[76-13-1] (1976)	1000	7670	1250	9590
Tricyclohexyltin hydroxide, *see* Cyhexatin					
Tridymite, *see* Silica—Crystaline					

(e) The value is for total dust containing no asbestos and < 1% crystalline silica.

(o) For control of general room air, biologic monitoring is essential for personnel control.

Capital letters A & B refer to Appendices; C denotes ceiling limit.

( ) Adopted values enclosed are on the Notice of Intended Changes list. Consult it for current proposal for change.

◄ Identifies substances for which there are also BEIs (*see* BEI section). Substances identified in the BEI documentations for methemoglobln inducers (for which methemoglobin is the principle toxicity) and organophosphorus cholinesterase inhibitors are part of this notation.

■ Substance identified by other sources as a suspected or confirmed human carcinogen. See the compliation in the Appendix to the Documentation of TLVs, pp. A-5(86)–A-9(86).

• Substance for which OSHA and/or NIOSH has a Permissible Exposure Limit (PEL) or a Recommended Exposure Limit (REL) lower than the TLV.

‡ See Notice of Intended Changes.

Substance	[CAS #]	TWA ppm[a]	TWA mg/m3[b]	STEL ppm[a]	STEL mg/m3[b]
Triethylamine [121-44-8] (1983)		10	41	15	62
Trifluorobromomethane [75-63-8] (1986)		1000	6090	—	—
Trimellitic anhydride [552-30-7] (1981)		0.005	0.059	—	—
Trimethylamine [75-50-3] (1983)		10	24	15	36
Trimethyl benzene [25551-13-7] (1987)		25	123	—	—
Trimethyl phosphite [121-45-9] (1986)		2	10	—	—
2,4,6-Trinitrophenol, *see* Picric acid					
2,4,6-Trinitrophenylmethylnitramine, *see* Tetryl					
▪◂2,4,6-Trinitrotoluene (TNT) [118-96-7] — Skin (1986)		—	0.5	—	—
Triorthocresyl phosphate [78-30-8] — Skin (1986)		—	0.1	—	—
Triphenyl amine [603-34-9] (1980)		—	5	—	—
Triphenyl phosphate [115-86-6] (1986)		—	3	—	—
Tripoli, *see* Silica — Crystaline					
Tungsten [7440-33-7], as W					
Insoluble compounds (1976)		—	5	—	10
Soluble compounds (1976)		—	1	—	3
Turpentine [8006-64-2] (1987)		100	556	—	—
•Uranium (natural) [7440-61-1] Soluble & insoluble compounds, as U (1976)		—	0.2	—	0.6
n-Valeraldehyde [110-62-3] (1978)		50	176	—	—
Vanadium, as V_2O_5 [1314-62-1] Respirable dust & fume (1982)		—	0.05	—	—
Vegetable oil mists[p] (1972)		—	10[e]	—	—
•Vinyl acetate [108-05-4] (1976)		10	35	20	70
Vinyl benzene, *see* Styrene					
•▪Vinyl bromide [593-60-2] (1980)		5,A2	22,A2	—	—
•▪Vinyl chloride [75-01-4] (1980)		5,A1	13,A1	—	—

Substance [CAS #]	ADOPTED VALUES TWA ppm[a)]	TWA mg/m³[b)]	STEL ppm[a)]	STEL mg/m³[b)]
Vinyl cyanide, *see* Acrylonitrile				
■Vinyl cyclohexene dioxide [106-87-6] — Skin (1977)	10,A2	57,A2	—	—
•■Vinylidene chloride [75-35-4] (1984)	5	20	20	79
Vinyl toluene [25013-15-4] (1981)	50	242	100	483
•VM & P Naphtha [8032-32-4] (1987)	300	1370	—	—
Warfarin [81-81-2] (1987)	—	0.1	—	—
Welding fumes (NOC†) (1977)	—	5,B2	—	—
■Wood dust (certain hard woods as beech & oak) (1981)	—	1	—	—
■ Soft wood (1981)	—	5	—	10
◄Xylene [1330-20-7; 95-47-6; 108-38-3; 106-42-3] (o-, m-, p-isomers) (1976)	100	434	150	651
m-Xylene α,α'-diamine [1477-55-0] — Skin (1977)	—	C 0.1	—	—
‡◄Xylidine [1300-73-8] — Skin	(2)	(9.9)	—	—
Yttrium [7440-65-5] metal & compounds, as Y (1988)	—	1	—	—

(e) The value is for total dust containing no asbestos and < 1% crystalline silica.
(p) Except caster, cashew nut, or similar irritant oils.
Capital letters A & B refer to Appendices; C denotes ceiling limit.
( ) Adopted values enclosed are on the Notice of Intended Changes list. Consult it for current proposal for change.
◄ Identifies substances for which there are also BEIs (*see* BEI section). Substances identified in the BEI documentations for methemoglobin inducers (for which methemoglobin is the principle toxicity) and organophosphorus cholinesterase inhibitors are part of this notation.
■ Substance identified by other sources as a suspected or confirmed human carcinogen. See the compliation in the Appendix to the Documentation of TLVs, pp. A-5(86)–A-9(86).
• Substance for which OSHA and/or NIOSH has a Permissible Exposure Limit (PEL) or a Recommended Exposure Limit (REL) lower than the TLV.
‡ See Notice of Intended Changes.
† NOC = not otherwise classified.

Substance [CAS #]	ADOPTED VALUES TWA ppm[a]	mg/m^{3}[b]	STEL ppm[a]	mg/m^{3}[b]
Zinc chloride fume [7646-85-7] (1976)	—	1	—	2
•■Zinc chromates [13530-65-9; 1103-86-9; 37300-23-5], as Cr (1988) ..	—	0.01,A1	—	—
Zinc oxide [1314-13-2]				
Fume (1976)	—	5	—	10
Dust (1976)	—	10[e]	—	—
Zirconium compounds [7440-67-7], as Zr (1976) ..	—	5	—	10

Dusts: See individual entries in the main alphabetical listing or Particulates Not Otherwise Classified.

Radioactivity: See Physical Agents section on Ionizing Radiation.

(e) The value is for total dust containing no asbestos and < 1% crystalline silica.
Capital letters A & B refer to Appendices; C denotes ceiling limit.

■ Substance identified by other sources as a suspected or confirmed human carcinogen. See the compliation in the Appendix to the Documentation of TLVs, pp. A-5(86)–A-9(86).

• Substance for which OSHA and/or NIOSH has a Permissible Exposure Limit (PEL) or a Recommended Exposure Limit (REL) lower than the TLV.

NOTICE OF INTENDED CHANGES (for 1989-90)

These substances, with their corresponding values, comprise those for which either a limit has been proposed for the first time, or for which a change in the "Adopted" listing has been proposed. In both cases, the proposed limits should be considered trial limits that will remain in the listing for a period of at least two years. If, after two years no evidence comes to light that questions the appropriateness of the values herein, the values will be reconsidered for the "Adopted" list. Documentation is available for each of these substances.

	TWA		STEL	
Substance [**CAS #**]	**ppm**[a]	**mg/m³**[b]	**ppm**[a]	**mg/m³**[b]
Acrylic acid [79-10-7] — Skin	2	5.9	—	—
Cadmium [7440-43-9] and compounds, as Cd	—	0.01,A2	—	—
†Calcium chromate [13765-19-0], as Cr	—	0.001,A2	—	—
†Calcium silicate (synthetic) [1344-95-2]	—	10	—	—
†Caprolactam [105-60-2]				
Dust	—	1	—	—
Vapor	5	23	—	—
†Chloroacetyl chloride [79-04-9] — Skin	0.05	0.23	0.15	0.66
†Chlorobenzene [108-90-7]	10	46	—	—
†2-Chloropropionic acid [598-78-7] — Skin	0.1	0.44	—	—
Dibutyl phenyl phosphate [2528-36-1] — Skin	0.3	3.5	—	—
†1,1-Dimethylhydrazine [57-14-7] — Skin	0.01,A2	0.025,A2	—	—
Ethyl acrylate [140-88-5]	5,A2	20,A2	15,A2	61,A2
†Formaldehyde [50-00-0]	C 0.3,A2	C 0.45,A2	—	—
Formic acid [64-18-6]	5	9.4	10	18

(a) Parts of vapor or gas per million parts of contaminated air by volume at 25°C and 760 torr.

(b) Milligrams of substance per cubic meter of air. When entry is in this column only, the value is exact; when listed with a ppm entry, it is approximate.

Capital letters A & B refer to Appendices; C denotes ceiling limit.

† 1989–1990 Revision or Addition.

Substance [CAS #]	TWA ppm[a)]	TWA mg/m³[b)]	STEL ppm[a)]	STEL mg/m³[b)]
†Graphite (all forms except graphite fibers)	—	2,[i)] Respirable dust		
†Hydrazine [302-01-2]—Skin	0.01,A2	0.013,A2	—	—
†Lead chromate [7758-97-6]				
as Pb	—	0.05,A2	—	—
as Cr	—	0.012,A2	—	—
†Methyl hydrazine [60-34-4]—Skin	0.01,A2	0.019,A2	—	—
†Morpholine [110-91-8]—Skin	20	71	—	—
†Nickel [7440-02-0]				
Metal	—	0.05,A1	—	—
Insoluble compounds, as Ni	—	0.05,A1	—	—
Soluble compounds, as Ni	—	0.05,A1	—	—
†Nickel carbonyl [13463-39-3], as Ni	Delete listing; nickel listing above applies.			
†Nickel sulfide roasting, fume & dust, as Ni	Delete listing; nickel listing above applies.			
†Pentachloronitrobenzene [82-68-8]	—	0.5	—	—
†m-Phenylenediamine [108-45-2]—Skin	—	0.1	—	—
†o-Phenylenediamine [95-54-5]	—	0.1,A2	—	—
†p-Phenylenediamine [106-50-3]	—	0.1	—	—
†Phenylhydrazine [100-63-0]—Skin	0.1,A2	0.44,A2	—	—
Picric acid [88-89-1]	—	0.1	—	—
†Silica fume (amorphous) [7631-86-9]	—	0.2,[i)]	Respirable dust	1[i)]
†Strontium chromate [7789-06-2], as Cr	—	0.001,A2	—	—
Thiram [137-26-8]	—	1	—	—
Tin, organic compounds, as Sn—Skin	—	0.1	—	0.2
Xylidine (mixed isomers) [1300-73-8] — Skin	0.5,A2	2.5,A2	—	—

DELETE* THE SHORT-TERM EXPOSURE LIMITS (TLVs-STELs) FOR THE FOLLOWING SUBSTANCES:

sec-Butyl alcohol
Chlordane
Chlorinated diphenyl oxide
Chlorobromomethane
Chlorodifluoromethane
Chlorodiphenyl (42% chlorine)
Chlorodiphenyl (54% chlorine)
Chloropicrin
o-Chlorotoluene
Chlorpyrifos
Clopidol
Crufomate
Cyclonite
Isobutyl acetate
Methyl acetylene
Phosphorus oxychloride
Picloram
Propionic acid

(j) These TLVs are for the respirable fraction of dust for the substance listed. The concentration of respirable dust for the application of this limit is to be determined from the fraction passing a size-selector with the characteristics defined in the "c." paragraphs of Appendix D.
Capital letters A & B refer to Appendices; C denotes ceiling limit.
† 1989–1990 Revision or Addition.

*As per the individual documentation appearing in the *Documentation of the Threshold Limit Values and Biological Exposure Indices,* 5th ed. (1986).

ADOPTED APPENDICES

APPENDIX A
Carcinogens

The Chemical Substances Threshold Limit Values Committee classifies certain substances found in the occupational environment as either confirmed or suspected human carcinogens. The present listing of substances which have been identified as carcinogens takes two forms: those for which a TLV has been assigned and those for which environmental and exposure conditions have not been sufficiently defined to assign a TLV. Where a TLV has been assigned, it does not necessarily imply the existence of a biological threshold; however, if exposures are controlled to this level, we would not expect to see a measurable increase in cancer incidence or mortality.

The TLV Committee considers information from the following kinds of studies to be indicators of a substance's potential to be a carcinogen in humans: epidemiology studies, toxicology studies, and, to a lesser extent, case histories. Scientific debate over the existence of biological thresholds for carcinogens is unlikely to be resolved in the near future. Because of the long latent period for many carcinogens, and for ethical reasons, it is often impossible to base timely risk-management decisions on results from human studies.

In order to recognize the qualitative differences in research results, two categories of carcinogens are designated in this booklet:

A1 — *Confirmed Human Carcinogens.* Substances, or substances associated with industrial processes, recognized to have carcinogenic potential.

and

A2 — *Suspected Human Carcinogens.* Chemical substances, or substances associated with industrial processes, which are suspect of inducing cancer, based on either limited epidemiological evidence or demonstration of carcinogenesis in one or more animal species by appropriate methods.

Exposures to carcinogens must be kept to a minimum. Workers exposed to A1 carcinogens without a TLV should be properly equipped to virtually eliminate all exposure to the carcinogen. For A1 carcinogens with a TLV and for A2 carcinogens, worker exposure by all routes should be carefully controlled to levels as low as reasonably achievable (ALARA) below the TLV. Please see the *Documentation of the Threshold Limit Values* for a more complete description and derivation of these designations.

APPENDIX B
Substances of Variable Composition

B1. *Polytetrafluoroethylene* decomposition products.* Thermal decomposition of the fluorocarbon chain in air leads to the formation of oxidized products containing carbon, fluorine, and oxygen. Because these products decompose in part by hydrolysis in alkaline solution, they can be quantitatively determined in air as fluoride to provide an index of exposure. No TLV is recommended pending determination of the toxicity of the products, but air concentration should be minimal.

* Trade Names: Algoflon, Fluon, Teflon, Tetran.

B2. ***Welding Fumes — Total Particulate (NOC)*****†**

TLV-TWA, 5 mg/m³

Welding fumes cannot be classified simply. The composition and quantity of both are dependent on the alloy being welded and the process and electrodes used. Reliable analysis of fumes cannot be made without considering the nature of the welding process and system being examined; reactive metals and alloys such as aluminum and titanium are arc-welded in a protective, inert atmosphere such as argon. These arcs create relatively little fume, but they do create an intense radiation which can produce ozone. Similar processes are used to arc-weld steels, also creating a relatively low level of fumes. Ferrous alloys also are arc-welded in oxidizing environments which generate considerable fume and can produce carbon monoxide instead of ozone. Such fumes generally are composed of discreet particles of amorphous slags containing iron, manganese, silicon, and other metallic constituents depending on the alloy system involved. Chromium and nickel compounds are found in fumes when stainless steels are arc-welded. Some coated and flux-cored electrodes are formulated with fluorides and the fumes associated with them can contain significantly more fluorides than oxides. Because of the above factors, arc-welding fumes frequently must be tested for individual constituents which are likely to be present to determine whether specific TLVs are exceeded. Conclusions based on total fume concentration are generally adequate if no toxic elements are present in welding rod, metal, or metal coating and conditions are not conducive to the formation of toxic gases.

Most welding, even with primitive ventilation, does not produce exposures inside the welding helmet above 5 mg/m³. That which does, should be controlled.

APPENDIX C
Threshold Limit Values for Mixtures

When two or more hazardous substances, which act upon the same organ system, are present, their combined effect, rather than that of either individually, should be given primary consideration. In the absence of information to the contrary, the effects of the different hazards should be considered as additive. That is, if the sum of the following fractions,

† Not otherwise classified (NOC).

$$\frac{C_1}{T_1} + \frac{C_2}{T_2} + \dots \frac{C_n}{T_n}$$

exceeds unity, then the threshold limit of the mixture should be considered as being exceeded. C_1 indicates the observed atmospheric concentration, and T_1 the corresponding threshold limit (*see* Example A.1 and B.1).

Exceptions to the above rule may be made when there is a good reason to believe that the chief effects of the different harmful substances are not in fact additive, but *independent* as when purely local effects on different organs of the body are produced by the various components of the mixture. In such cases the threshold limit ordinarily is exceeded only when at least one member of the series (C_1/T_1 + or + C_2/T_2, etc.) itself has a value exceeding unity (*see* Example B.1).

Synergistic action or potentiation may occur with some combinations of atmospheric contaminants. Such cases at present must be determined individually. Potentiating or synergistic agents are not necessarily harmful by themselves. Potentiating effects of exposure to such agents by routes other than that of inhalation is also possible, e.g., imbibed alcohol and inhaled narcotic (trichloroethylene). Potentiation is characteristically exhibited at high concentrations, less probably at low.

When a given operation or process characteristically emits a number of harmful dusts, fumes, vapors or gases, it will frequently be only feasible to attempt to evaluate the hazard by measurement of a single substance. In such cases, the threshold limit used for this substance should be reduced by a suitable factor, the magnitude of which will depend on the number, toxicity, and relative quantity of the other contaminants ordinarily present.

Examples of processes which are typically associated with two or more harmful atmospheric contaminants are welding, automobile repair, blasting, painting, lacquering, certain foundry operations, diesel exhausts, etc.

Examples of TLVs for Mixtures

A. *Additive effects.* The following formulae apply only when the components in a mixture have similar toxicologic effects; they should not be used for mixtures with widely differing reactivities, e.g., hydrogen cyanide and sulfur dioxide. In such case, the formula for Independent Effects (B) should be used.

1. General case, where air is analyzed for each component, the TLV of mixture =

$$\frac{C_1}{T_1} + \frac{C_2}{T_2} + \frac{C_3}{T_3} + \dots = 1$$

Note: It is essential that the atmosphere be analyzed both qualitatively and quantitatively for each component present in order to evaluate compliance or noncompliance with this calculated TLV.

Example A.1: Air contains 400 ppm of acetone (TLV, 750 ppm), 150 ppm of sec-butyl acetate (TLV, 200 ppm) and 100 ppm of methyl ethel ketone (TLV, 200 ppm).

Atmospheric concentration of mixture = 400 + 150 + 100 = 650 ppm of mixture.

$$\frac{400}{750} + \frac{150}{200} + \frac{100}{200} = 0.53 + 0.75 + 0.5 = 1.78$$

Threshold Limit is exceeded.

2. Special case when the source of contaminant is a liquid mixture and the atmospheric composition is *assumed* to be similar to that of the original material, e.g., on a time-weighted average exposure basis, all of the liquid (solvent) mixture eventually evaporates. When the percent composition (by weight) of the liquid mixture is known, the TLVs of the constituents must be listed in mg/m³. TLV of mixture =

$$\frac{1}{\dfrac{f_a}{TLV_a} + \dfrac{f_b}{TLV_b} + \dfrac{f_c}{TLV_c} + \ldots \dfrac{f_n}{TLV_n}}$$

Note: In order to evaluate compliance with this TLV, field sampling instruments should be calibrated, in the laboratory, for response to this specific quantitative and qualitative air-vapor mixture, and also to fractional concentrations of this mixture, e.g., 1/2 the TLV; 1/10 the TLV; 2 × the TLV; 10 × the TLV; etc.)

Example A.2: Liquid contains (by weight):

50% heptane: TLV = 400 ppm or 1600 mg/m³
1 mg/m³ ≡ 0.25 ppm

30% methyl chloroform: TLV = 350 ppm or 1900 mg/m³
1 mg/m³ ≡ 0.18 ppm

20% perchloroethylene: TLV = 50 ppm or 335 mg/m³
1 mg/m³ ≡ 0.15 ppm

$$\text{TLV of Mixture} = \frac{1}{\dfrac{0.5}{1600} + \dfrac{0.3}{1900} + \dfrac{0.2}{335}}$$

$$= \frac{1}{0.00031 + 0.00016 + 0.0006}$$

$$= \frac{1}{0.00107} = 935 \text{ mg/m}^3$$

of this mixture

50% or (935)(0.5) = 468 mg/m^3 is heptane
30% or (935)(0.3) = 281 mg/m^3 is methyl chloroform
20% or (935)(0.2) = 187 mg/m^3 is perchloroethylene

These values can be converted to ppm as follows:

heptane: 468 mg/m^3 × 0.25 = 117 ppm
methyl chloroform: 281 mg/m^3 × 0.18 = 51 ppm
perchloroethylene: 187 mg/m^3 × 0.15 = 29 ppm

TLV of mixture = 117 + 51 + 29 = 197 ppm, or 935 mg/m^3

B. *Independent effects.* TLV for mixture =

$$\frac{C_1}{T_1} = 1; \quad \frac{C_2}{T_2} = 1 \quad \frac{C_3}{T_3} = 1; \text{ etc.}$$

Example B.1: Air contains 0.15 mg/m^3 of lead (TLV, 0.15) and 0.7 mg/m^3 of sulfuric acid (TLV, 1).

$$\frac{0.15}{0.15} = 1; \qquad \frac{0.7}{1} = 0.7$$

Threshold limit is not exceeded.

C. *TLV for mixtures of mineral dusts.* For mixtures of biologically active mineral dusts the general formula for mixtures given in A.2 may be used.

APPENDIX D

Particle Size-Selective Sampling Criteria for Airborne Particulate Matter

For chemical substances present in inhaled air as suspensions of solid particles or droplets, the potential hazard depends on particle size as well as mass concentration because of: 1) effects of particle size on deposition site within the respiratory tract, and 2) the tendency for many occupational diseases to be associated with material deposited in particular regions of the respiratory tract.

ACGIH has recommended particle size-selective TLVs for crystalline silica for many years in recognition of the well established association between silicosis and respirable mass concentrations. It now has embarked on a re-examination of other chemical substances encountered in particulate form in occupational environments with the objective of defining: 1) the size-fraction most closely associated for each substance with the health effect of concern, and 2) the mass concentration within that size fraction which should represent the TLV.

The Particle Size-Selective TLVs (PSS-TLVs) will be expressed in three forms, e.g.,

a. *Inspirable Particulate Mass TLVs (IPM-TLVs)* for those materials which are hazardous when deposited anywhere in the respiratory tract.
b. *Thoracic Particulate Mass TLVs (TPM-TLVs)* for those materials which are hazardous when deposited anywhere within the lung airways and the gas-exchange region.
c. *Respirable Particulate Mass TLVs (RPM-TLVs)* for those materials which are hazardous when deposited in the gas-exchange region.

The three particulate mass fractions described above are defined in quantitative terms as follows:

a. Inspirable Particulate Mass consists of those particles that are captured according to the following collection efficiency regardless of sampler orientation with respect to wind direction:

$$E = 50(1 + \exp[-0.06\ d_a]) \pm 10;$$
$$\text{for } 0 < d_a \leqslant E\ 100\ \mu m$$

   Collection characteristics for $d_a > 100\ \mu m$ are presently unknown. E is collection efficiency in percent and d_a is aerodynamic diameter in μm.
b. Thoracic Particulate Mass consists of those particles that penetrate a separator whose size collection efficiency is described by a cumulative lognormal function with a median aerodynamic diameter of 10 $\mu m \pm 1.0\ \mu m$ and with a geometric standard deviation of 1.5 (± 0.1).
c. Respirable Particulate Mass consists of those particles that penetrate a separator whose size collection efficiency is described by a cumulative lognormal function with a median aerodynamic diameter of 3.5 $\mu m \pm 0.3\ \mu m$ and with a geometric standard deviation of 1.5 (± 0.1). This incorporates and clarifies the previous ACGIH Respirable Dust Sampling Criteria.

These definitions provide a range of acceptable performance for each type of size-selective sampler. Further information is available on the background and performance criteria for these particle size-selective sampling recommendations.[1]

References

1. **ACGIH:** *Particle Size-Selective Sampling in the Workplace,* 80 pp. Cincinnati, Ohio (1984).

APPENDIX E
Chemical Substances and Other Issues Under Study

Information, data especially, and comments are solicited to assist the Committee in its deliberations and in the possible development of draft documents. Draft documentations are used by the Committee to decide what action, if any, to recommend on a given question.

Chemical Substances

Acetaldehyde
Acetomethylchloride
Acetophenone
Acrylic acid
Adipic acid
Adiponitrile
Bentonite
Benzyl acetate
Borax and boron compounds
Bromine pentafluoride
Bromochloromethane
Bromodichloromethane
Bromoform
n-Butyl acetate
2-t-Butylazo-2-hydroxy-5-methylhexane
Cadmium
Carbon disulfide
Carbon monoxide
Chromium
Chrysene
Dibutyl phenyl phosphate
Dichlorodiphenyl sulfone
Dichlorocyclopentadiene
2,4-D (2,4-Dichlorophenoxy acetic acid)
1,3-Dichloropropene
Dichlorvos
Diethylamine
Dimethylamine
Dimethyl acetamide
Dimethyl formamide
Dimethyl disulfide
Dinitrotoluene
Epichlorohydrin
EPN
Ethyl bromide
2-Ethyl hexanoic acid
Gallium arsenide
Gasoline (unleaded)
Glutaraldehyde
Glycol ethers
Graphite fibers
Heptachlor
Hexachlorobenzene
Hexachlorocyclopentadiene
Hexamethylene diamine
Jet, petroleum and diesel fuels
Inorganic lead
Malathion
Man-made mineral fibers
2-Methoxyethanol
Methyl bromide
Methylene diamine
4,4′-Methylene dianiline
Methyl tert-butyl ether
Mineral spirits
Naled
Nitromethane
Pentachlorophenol
2,4-Pentanedione
Perchloroethylene
Perfluoroisobutylene
Persulfates
Petroleum solvents
Propylene dichloride
Sulfur tetrafluoride
Sulfuryl fluoride
Tantalum
Terphenyls

Terephthalic acid
1,1,1,2-Tetrachloro-2,2-difluoroethane
1,1,2,2-Tetrachloro-1,2-difluoroethane
1,1,2,2-Tetrachloroethane
Tetrakis phosphonium chloride
Tetrakis phosphonium sulfate
Tetrasodium pyrophosphate
Tobacco smoke
Trichloroethylene
Triethanolamine
Triethylamine
Vinyl cyclohexene
Uranium

Other Issues

1. Solubility.
2. Should the TLVs currently expressed as "total dust" be changed to "inspirable particulate mass," as defined in Appendix D, without changing the numerical value?
3. Two working groups have been formed to address questions of:
   a. Altered work shifts.
   b. Skin notation criteria.
4. Excursion limits

Biological Exposure Indices

Adopted by ACGIH with Intended Changes for 1989–90

1988–89 BIOLOGICAL EXPOSURE INDICES COMMITTEE

Vera F. Thomas, Ph.D., University of Miami—*Chair*
Pierre O. Droz, Ph.D., Lausanne University, Switzerland
Philip A. Edelman, M.D., University of California—Irvine
Larry K. Lowry, Ph.D., NIOSH
Masana Ogata, M.D., Okayama University Medical School, Japan
J. Thomas Pierce, Ph.D., Virginia Commonwealth University
Jon Rosenberg, M.D., California Dept. of Health Services/Dept. of Industrial Relations
Janice W. Yager, Ph.D., University of California—Berkeley

CONSULTANTS

Gerald L. Kennedy, Jr., Haskell Laboratory, E.I. du Pont de Nemours & Co., Inc.
Mitchell R. Zavon, M.D., Agatha Corporation

INTRODUCTION TO THE BIOLOGICAL EXPOSURE INDICES

Biological monitoring provides occupational health personnel with a tool for assessing worker's exposure to chemicals. ***Workplace air monitoring*** consists of assessment of inhalation exposure to chemicals in the workplace through measurement of the chemical concentration in the ambient air. ***TLVs*** serve as a reference value. ***Biological monitoring*** consists of an assessment of overall exposure to chemicals which are present in the workplace through measurement of the appropriate determinant(s) in biological specimens collected from the worker at the specified time. ***BEIs*** (Biological Exposure Indices) serve as a reference value.

The ***determinant*** can be the chemical itself or its metabolite(s), or a characteristic reversible biochemical change induced by the chemical. The measurement can be made in exhaled air, urine, blood, or other biological specimens collected from the exposed worker. Based on the determinant, the specimen chosen, and the time of sampling, the measurement indicates either the intensity of a recent exposure, an average daily exposure, or a chronic cumulative exposure.

Biological Exposure Indices (BEIs) are reference values intended as guidelines for the evaluation of potential health hazards in the practice of industrial hygiene. BEIs represent the levels of determinants which are most likely to be observed in specimens collected from a healthy worker who has been exposed to chemicals to the same extent as a worker with inhalation exposure to the TLV-TWA. BEIs do not indicate a sharp distinction between hazardous and nonhazardous exposures. Due to biological variability it is possible for an individual's measurements to exceed the BEI without incurring an increased health risk. If, however, measurements in specimens obtained from a worker on different occasions persistently exceed the BEI, or if the majority of measurements in specimens obtained from a group of workers at the same workplace exceed the BEI, the cause of the excessive values must be investigated and proper action taken to reduce the exposure.

BEIs apply to eight-hour exposures, five days a week. However, BEIs for altered working schedules can be extrapolated on pharmacokinetic and pharmacodynamic bases. BEIs should not be applied either directly or through a conversion factor, in the determination of safe levels for nonoccupational exposure to air and water pollutants, or food contaminants. The BEIs are not intended for use as a measure of adverse effects or for diagnosis of occupational illness.

The data base for each BEI recommendation consists of available information on absorption, elimination, and metabolism of chemicals, and on the correlation between exposure intensity and biological effect in workers. The BEI is based either on the relationship between intensity of exposure and biological levels of the determinant or on the relationship between biological levels and health effects. Human data from controlled and field studies are used to find such relationships. Animal studies usually do not provide data suitable for establishing a BEI; however, since the BEIs are usually related to exposures to TLV, the BEI is indirectly based on the dose-response animal studies used in establishing the TLV.

Implementation: Biological monitoring should be considered complementary to air monitoring. It should be conducted when it offers an advantage over the use of air monitoring alone. Biological monitoring should be used to substantiate air monitoring, to test the efficacy of personal protective equipment, to determine the potential for absorption via the skin and the gastrointestinal system, or to detect nonoccupational exposure. The existence of a BEI does not indicate the need for conducting biological monitoring. Occupational health personnel must exercise professional judgment in designing monitoring protocols. The documentations are intended to provide helpful background information.

Interpretation of data: When biological monitoring data are interpreted, intraindividual and interindividual differences in tissue levels of determinants occurring at the same exposure conditions must be considered. Such differences arise on account of variation in pulmonary ventilation, hemodynamics, body composition, efficacy of excretory organs, and activity of enzyme systems which mediate metabolism of the chemical. Multiple sampling is necessary to reduce the effects of variable factors. Biological monitoring may confirm the results of air monitoring, but where there is a discrepancy between the results, the entire exposure situation should be carefully reviewed and an explanation found.

The main source of inconsistency in information on exposure intensity by air monitoring and biological monitoring is the variability of the following factors: ***physiological and health status of the worker***, such as body build, diet (water and fat intake), enzymatic activity, body fluid composition, age, sex, pregnancy, medication, and disease state; occupational ***exposure sources***, such as the intensity of the physical work load and fluctuation of exposure intensity, skin exposure, temperature and humidity, coexposure to other chemicals; ***environmental sources***, such as community and home air pollutants, and water and food contaminants; ***individual life style sources***, such as after work activities, personal hygiene, working and eating habits, smoking, alcohol and drug intake, exposure to household products, or exposure to chemicals from hobbies or from another workplace; and ***methodological sources***, which include specimen contamination and deterioration during collection, storage and analysis, and bias of the selected analytical methods. The significance of these effects must be assessed individually for each situation. Drugs, pollutants or coexposure to another chemical can alter the relationship between the intensity of occupational exposure and the level of the determinant in the specimen either by adding to the level of the determinant or by altering the metabolism or elimination of the studied chemical. The documentations provide specific information on the effects of these factors.

The timing indicates when the sample should be collected with respect to the exposure. It must be carefully observed because distribution and elimination of a chemical or its metabolic products, as well as biochemical changes induced by exposure to the chemical, are kinetic events. The listed BEIs are applicable only if collection is conducted at the specified time.

The quality control program of the laboratory performance is essential to reduce analytical errors and bias in the results.

Comments on the table of BEIs: The table specifies, for indicated chemicals, the determinant, the specimen to be collected, the time at which the specimen must be collected, and the BEI. Additional important information is indicated by notation(s).

Determinants: The table comprises BEIs for all determinants where a sufficient data base is available and on which the BEI Committee took action. Occupational health personnel must use professional judgment to decide which determinant should be used in specific instances in order to meet the objectives of the monitoring.

Biological specimens: Urine, exhaled air, and blood specimens are recommended. Each type of specimen has distinctive causes of variability which affect the level of the determinant in the specimen. Other specimens, such as hair or nails, are not recommended at this time.

For data based on ***urine analysis***, variation in urine volume is the most significant. Measurements of elimination rate usually provide more precise information. However, quantitative collection of urine during a precise time period is rarely possible. A simple measurement of concentration can provide information on exposure, but the quantitative measure of exposure is weakened by the variability of rate of urine output. The urinary concentration related to excretion of the solute provides some correction for fluctuation of urine output. Highly diluted and highly concentrated urine specimens are usually not suitable for monitoring, and a new specimen should be taken.

For data based on ***exhaled air analysis***, the rapid changes of the concentration with time are critical; moreover, the concentration changes during the expiration phase. Therefore, the sampling of end-exhaled air (which usually represents alveolar air) or mixed-exhaled air is specified. In general, during exposure the concentrations in end-exhaled air are smaller than in mixed-exhaled air, and during postexposure the concentration in mixed-exhaled air accounts for about two-thirds of the concentration in end-exhaled air. Exhaled air specimens collected from workers with altered pulmonary function may not be suitable for exposure monitoring.

For data based on ***blood analysis***, the plasma-erythrocyte ratio and the distribution of some determinants among blood constituents can affect the outcome of some measurements. Therefore, the analysis of whole blood, plasma, serum, or erythrocytes is specified. The protein binding of some determinants should be considered when the analytical method is selected. The concentration difference between arterial and venous blood induced by pulmonary uptake or pulmonary clearance has to be considered when blood is collected for measurements of volatile chemicals. Unless otherwise indicated, the BEIs for volatile chemicals relate to venous blood and cannot be applied to capillary blood, which mainly represents arterial blood.

Sampling time: In many instances, when the level of the determinant changes rapidly or when accumulation occurs, the sampling time is very critical and must be carefully observed. The sampling time is specified in the table according to the differences in the uptake and elimination rates of chemicals and their metabolites, and according to the persistence of induced biochemical changes, as follows:

1. Determinants with timing "prior to shift" (meaning after 16 hours without exposure), "during shift," or "end of shift" (meaning the last two hours of exposure) are eliminated rapidly with a half-time less than five hours. Such determinants do not accumulate in the body and, therefore, their timing is critical only in relation to the exposure and postexposure periods.
2. Determinants with timing "beginning of workweek," or "end of workweek" (meaning after two days without exposure or after four or five consecutive working days with exposure, respectively) are eliminated with half-times longer than five hours. Such determinants accumulate in the body during the workweek; therefore, their timing is critical in relation to previous exposures. For chemicals with multiphase elimination, the timing is given in relation to the workday exposure (shift) as well as to the workweek exposure.
3. Determinants with timing "not critical," or "discretionary," have very long elimination half-times and accumulate in the body over years, some for a lifetime. After a couple of weeks of exposure, specimens for measurements of such determinants can be collected any time.

Notations provide the following information:

*"**Sc**" **Notation***. This notation indicates that an identifiable population group might have an increased ***susceptibility*** to the effect of the chemical, thus leaving it unprotected by the recommended BEI. The specific documentation should be consulted for information.

*"**B**" **Notation***. This notation indicates that the determinant is usually present in a significant amount in biological specimens collected from subjects who have not been occupationally exposed. Such background levels are included in the BEI value. For information on ***background levels,*** consult the specific documentation.

*"**Ns**" **Notation***. This notation indicates that the determinant is ***nonspecific***, since it is observed after exposure to some other chemicals. These nonspecific tests are preferred because they are easy to use and usually offer a better correlation with exposure than specific tests. In such instances, a BEI for a specific, less quantitative biological determinant is recommended as a confirmatory test. The documentation should be consulted for information on factors affecting interpretation of these BEIs.

*"**Cf**" **Notation***. This notation indicates that the biological determinant is an indicator of exposure to the chemical, but the quantitative interpretation of the measurements is ambiguous. Their BEIs should be applied cautiously. These biological determinants should be used as ***confirmatory*** tests, mainly for confirmation of exposures indicated by measurements of a nonspecific determinant or as a screening test if a quantitative test is not practical.

It is strongly advisable to consult the specific documentation published in the *Documentation of Threshold Limit Values and Biological Exposure Indices,* 5th edition, before designing biological monitoring and interpreting BEIs. Action on unexpected values should not be based on a single isolated measurement but on measurements of multiple sampling.

ADOPTED BIOLOGICAL EXPOSURE DETERMINANTS

Airborne Chemical [CAS #] *Indices*	*Sampling Time*	*BEI*	*Notation*
BENZENE [71-43-2]			
Total phenol in urine	End of shift	50 mg/L	B, Ns
Benzene in exhaled air:	Prior to next shift		
mixed-exhaled		0.08 ppm	Cf
end-exhaled		0.12 ppm	Cf
CADMIUM [7440-43-9]			
Cadmium in urine	Not critical	10 μg/g creatinine	B
Cadmium in blood	Not critical	10 μg/L	B
CARBON DISULFIDE [75-15-0]			
2-Thiothiazolidine-4-carboxylic acid (=TTCA) in urine	End of shift	5 mg/g creatinine	
CARBON MONOXIDE [630-08-0]			Sc
Carboxyhemoglobin in blood	End of shift	less than 8%	B, Ns
CO in end-exhaled air	End of shift	less than 40 ppm	B, Ns
DIMETHYLFORMAMIDE [68-12-2]			
N-Methylformamide in urine	End of shift	40 mg/g creatine	
ETHYL BENZENE [100-41-4]			
Mandelic acid in urine	End of shift at end of workweek	2 g/L 1.5 g/g creatinine	Ns

Airborne Chemical [CAS #] Indices	Sampling Time	BEI	Notation
Ethyl benzene in end-exhaled air	Prior to next shift	2 ppm	Cf
n-HEXANE [110-54-3]			
2,5-Hexanedione in urine	End of shift	5 mg/L	Ns
n-Hexane in end-exhaled air	During shift	40 ppm	Cf
LEAD [7439-92-1]			Sc
Lead in blood	Not critical	50 μg/100 ml	B
Lead in urine	Not critical	150 μg/g creatinine	B
Zinc protoporphyrin in blood	After 1 month exposure	250 μg/100 ml erythrocytes or 100 μg/100 ml blood	B
*METHYL CHLOROFORM [71-55-6]			
Methyl chloroform in end-exhaled air	Prior to the last shift of workweek	40 ppm	
Trichloroacetic acid in urine	End of workweek	10 mg/L	Ns, Cf
Total trichloroethanol in urine	End of shift at end of workweek	30 mg/L	Ns, Cf
Total trichloroethanol in blood	End of shift at end of workweek	1 mg/L	Ns
METHYL ETHYL KETONE (MEK) [78-93-3]			
MEK in urine	End of shift	2 mg/L	
*ORGANOPHOSPHORUS CHOLINESTERASE INHIBITORS			
Cholinesterase activity in red cells	Discretionary	70% of individual's baseline	B, Ns, Cf

* 1989–1990 Adoption.

Airborne Chemical [CAS #]

Indices	*Sampling Time*	*BEI*	*Notation*
*PARATHION [56-38-2]			
Total p-nitrophenol in urine	End of shift	0.5 mg/L	Ns, Cf
Cholinesterase activity in red cells	Discretionary	70% of individual's baseline	B, Ns, Cf
PENTACHLOROPHENOL (PCP) [87-86-5]			
Total PCP In urine	Prior to the last shift of workweek	2 mg/L	B
Free PCP in plasma	End of shift	5 mg/L	B
*PERCHLOROETHYLENE [127-18-4]			
Perchloroethylene in end-exhaled air	Prior to the last shift of workweek	10 ppm	
Perchloroethylene in blood	Prior to the last shift of workweek	1 mg/L	
Trichloroacetic acid in urine	End of workweek	7 mg/L	Ns, Cf
PHENOL [108-95-2]			
Total phenol in urine	End of shift	250 mg/g creatinine or 15 mg/hr	B, Ns
STYRENE [100-42-5]			
Mandelic acid in urine	End of shift	1 g/L 0.8 g/g creatinine	Ns Ns
Styrene in mixed-exhaled air	Prior to next shift	40 ppb	Cf
Phenylglyoxylic acid in urine	End of shift	250 mg/L 240 mg/g creatinine	B, Ns

* 1989–1990 Adoption.

Airborne Chemical [CAS #]

Indices	*Sampling Time*	*BEI*	*Notation*
Styrene in mixed-exhaled air	During shift	18 ppm	Cf
Styrene in blood	End of shift	0.55 mg/L	Cf
	Prior to next shift	0.02 mg/L	Cf
TOLUENE [108-88-3]			
Hippuric acid in urine	End of shift	2.5 g/g creatinine	B, Ns
	Last 4 hrs of shift	3 mg/min	
Toluene in venous blood	End of shift	1 mg/L	Cf
Toluene in end-exhaled air	During shift	20 ppm	Cf
TRICHLOROETHYLENE [79-01-6]			
Trichloroacetic acid in urine	End of workweek	100 mg/L	Ns
Trichloroacetic acid and trichloroethanol in urine	End of shift at end of workweek	300 mg/L	Ns
		320 mg/g creatinine	
Free trichloroethanol in blood	End of shift at end of workweek	4 mg/L	Ns
Trichloroethylene in end-exhaled air	Prior to the last shift of workweek	0.5 ppm	Cf
XYLENES [1330-20-7]			
Methylhippuric acids in urine	End of shift	1.5 g/g creatinine	
	Last 4 hrs of shift	2 mg/min	

NOTICE OF INTENT TO ESTABLISH

Airborne Chemical [CAS #] *Indices*	*Sampling Time*	*BEI*	*Notation*
†ANILINE [62-53-3]			
Total p-aminophenol in urine	End of shift	50 mg/g creatinine	Ns
Methemoglobin	End of shift	1.5%	Ns Cf
CHROMIUM (VI), Water Soluble Fume			
Total chromium in urine	Increase during shift	10 μg/g creatinine	B
	End of shift at end of workweek	30 μg/g creatinine	B
FLUORIDES			
Fluoride in urine	Prior to shift	3 mg/L	B, Ns
	End of shift	10 mg/L	B, Ns
†FURFURAL [98-01-01]			
Total furoic acid in urine	End of Shift	200 mg/g creatinine	B· Ns
†METHANOL [67-56-1)			
Methanol in urine	End of Shift	15 mg/L	B, Ns
Formic acid in urine	Before the shift at end of workweek	80 mg/g creatinine	B,Ns

†1989–1990 Addition or Revision.

Airborne Chemical [CAS #] *Indices*	*Sampling Time*	*BEI*	*Notation*
METHEMOGLOBIN INDUCERS			
Methemoglobin in blood	During or end of shift	1.5%	B, Ns, Cf
†NITROBENZENE [98-95-3]			
Total p-nitrophenol in urine	End of shift at end of workweek	5 mg/g creatinine	Ns
Methemoglobin	End of shift	1.5%	Ns, Cf

†1989–1990 Addition or Revision.

CHEMICAL SUBSTANCES UNDER STUDY TO ESTABLISH BIOLOGICAL EXPOSURE INDICES

Antimony
Arsenic
Benzene
Cadmium
Chlorobenzene
2-Ethoxyethanol
2-Ethoxyethyl acetate
Mercury
2-Methoxyethanol
2-Methoxyethyl acetate
Methylene chloride
Polychlorinated biphenyls
Selenium
Synthetic pyrethrins
Uranium
Vanadium

Threshold Limit Values for Physical Agents in the Work Environment

Adopted by ACGIH with Intended Changes for 1989–90

1988–89 PHYSICAL AGENTS TLV COMMITTEE

David H. Sliney, U.S. Army Environmental Hygiene Agency—*Chair*
Peter A. Breysse, University of Washington
Irving H. Davis, Retired
Zory R. Glaser, Ph.D., National Center for Medical Devices & Radiological Health/FDA
Allan P. Heins, Ph.D., OSHA—*Secretary*
Herbert H. Jones, Retired
Edward J. Largent, Retired
John C. Mitchell, USAF
Deirdre A. Morison, P.Eng., National Health and Welfare, Canada
Anthony M. Muc, Ph.D., Ontario Hydro
William E. Murray, NIOSH
Wordie H. Parr, Ph.D., Retired
Thomas K. Wilkinson, Retired

CONSULTANTS

Gerald V. Coles, Retired
Richard A. Ilka, M.D., Marathon Oil Company
Robert N. Ligo, M.D., E.I. du Pont de Nemours & Company, Inc.
Donald E. Wasserman, Vibration Consultant

INTRODUCTION TO THE PHYSICAL AGENTS*

These Threshold Limit Values (TLVs) refer to levels of physical agents and represent conditions under which it is believed that nearly all workers may be repeatedly exposed day after day without adverse effect. Because of wide variations in individual susceptibility, exposure of an occasional individual at, or even below, the threshold limit may not prevent annoyance, aggravation of a pre-existing condition, or physiological damage.

These threshold limits are based on the best available information from industrial experience, from experimental human and animal studies, and when possible, from a combination of the three.

These limits are intended for use in the practice of industrial hygiene and should be interpreted and applied only by a person trained in this discipline. They are not intended for use, or for modification for use, 1) in the evaluation or control of the levels of physical agents in the community, 2) as proof or disproof of an existing physical disability, or 3) for adoption by countries whose working conditions differ from those in the United States of America.

These values are reviewed annually by the Committee on Threshold Limits for Physical Agents for revision or additions, as further information becomes available.

The ACGIH disclaims any liability with respect to the use of TLVs.

Notice of Intent — At the beginning of each year, proposed actions of the Committee for the forthcoming year are issued in the form of a "Notice of Intent." This notice provides not only an opportunity for comment but also solicits suggestions of physical agents to be added to the list. The suggestions should be accompanied by substantiating evidence.

Definitions — Categories of Threshold Limit Values (TLVs) are specified herein, as follows:

a) *Threshold Limit Value–Time-Weighted Average (TLV–TWA)* — the time-weighted average exposure for a normal 8-hour workday and a 40-hour workweek, to which nearly all workers may be repeatedly exposed day after day, without adverse effect. Examples of their use can be found in the TLVs for Heat and Noise.

b) *Threshold Limit Value–Ceiling (TLV–C)* — the exposure that should not be exceeded even instantaneously, as in the case of 115 dBA limit for noise.

---

*Major text changes are indicated by a vertical rule in the margin.

Physical Factors. It is recognized that combinations of such physical factors as heat, ultraviolet and ionizing radiation, humidity, abnormal pressure (altitude), and the like may place added stress on the body so that the effects from exposure at a threshold limit may be altered. Also, most of these stresses may act adversely to increase the toxic response of a foreign substance. Although most threshold limits have built-in safety factors to guard against adverse effects to moderate deviations from normal environments, the safety factors of most exposures are not of such a magnitude as to take care of gross deviations. For example, continuous work at WBGT temperatures above 30°C (86°F), or overtime extending the workweek more than 25%, might be considered gross deviations. In such instances, judgment must be exercised in the proper adjustments of the TLVs.

ADOPTED THRESHOLD LIMIT VALUES

HEAT STRESS

These Threshold Limit Values (TLVs) refer to heat stress conditions under which it is believed that nearly all workers may be repeatedly exposed without adverse health effects. The TLVs shown in Table 1 are based on the assumption that nearly all acclimatized, fully clothed workers with adequate water and salt intake should be able to function effectively under the given working conditions without exceeding a deep body temperature of 38°C.[1,2]

Since measurement of deep body temperature is impractical for monitoring the workers' heat load, the measurement of environmental factors is required which most nearly correlate with deep body temperature and other physiological responses to heat. At the present time Wet Bulb Globe Temperature Index (WBGT) is the simplest and most suitable technique to measure the environmental factors. WBGT values are calculated by the following equations:

1. Outdoors with solar load:

   $$WBGT = 0.7\ NWB + 0.2\ GT + 0.1\ DB$$

2. Indoors or Outdoors with no solar load:

   $$WBGT = 0.7\ NWB + 0.3\ GT$$

where:

WBGT = Wet Bulb Globe Temperature Index
NWB = Natural Wet-Bulb Temperature
DB = Dry-Bulb Temperature
GT = Globe Temperature

The determination of WBGT requires the use of a black globe thermometer, a natural (static) wet-bulb thermometer, and a dry-bulb thermometer.

Higher heat exposures than shown in Table 1 are permissible if the workers have been undergoing medical surveillance and it has been established that they are more tolerant to work in heat than the average worker. Workers should not be permitted to continue their work when their deep body temperature exceeds 38.0°C.

TABLE 1
Permissible Heat Exposure Threshold Limit Values
(Values are given in °C WBGT)

	Work Load		
Work—Rest Regimen	**Light**	**Moderate**	**Heavy**
Continuous work	30.0	26.7	25.0
75% Work — 25% Rest, each hour	30.6	28.0	25.9
50% Work — 50% Rest, each hour	31.4	29.4	27.9
25% Work — 75% Rest, each hour	32.2	31.1	30.0

Evaluation and Control

I. *Measurement of the Environment*

The instruments required are a dry-bulb, a natural wet-bulb, a globe thermometer, and a stand. The measurement of the environmental factors shall be performed as follows:

A. The range of the dry and the natural wet bulb thermometer shall be –5°C to 50°C with an accuracy of ± 0.5°C. The dry bulb thermometer must be shielded from the sun and the other radiant surfaces of the environment without restricting the airflow around the

TABLE 2
Assessment of Work Load[9]

Average values of metabolic rate during different activities.

A. Body position and movement	**kcal/min**
Sitting	0.3
Standing	0.6
Walking	2.0-3.0
Walking up hill	add 0.8 per meter (yard) rise

B. Type of Work		**Average kcal/min**	**Range kcal/min**
Hand work	*light*	0.4	0.2-1.2
	heavy	0.9	
Work with one arm	*light*	1.0	0.7-2.5
	heavy	1.7	
Work with both arms	*light*	1.5	1.0-3.5
	heavy	2.5	
Work with body	*light*	3.5	2.5-15.0
	moderate	5.0	
	heavy	7.0	
	very heavy	9.0	

bulb. The wick of the natural wet-bulb thermometer shall be kept wet with distilled water for at least 1/2 hour before the temperature reading is made. It is not enough to immerse the other end of the wick into a reservoir of distilled water and wait until the whole wick becomes wet by capillarity. The wick shall be wetted by direct application of water from a syringe 1/2 hour before each reading. The wick shall extend over the bulb of the thermometer, covering the stem about one additional bulb length. The wick should always be clean and new wicks should be washed before using.

B. A globe thermometer, consisting of a 15-cm (6-inch) diameter hollow copper sphere painted on the outside with a matte black finish or equivalent, shall be used. The bulb or sensor of a thermometer (range –5° to +100°C with an accuracy of ± 0.5°C) must be fixed in the center of the sphere. The globe thermometer shall be exposed at least 25 minutes before it is read.

C. A stand shall be used to suspend the three thermometers so that they do not restrict free air flow around the bulbs, and the wet-bulb and globe thermometer are not shaded.

D. It is permissible to use any other type of temperature sensor that gives identical reading as that of a mercury thermometer under the same conditions.

TABLE 3

Activity Examples[9]

- Light hand work: writing, hand knitting
- Heavy hand work: typewriting
- Heavy work with one arm: hammering in nails (shoemaker, upholsterer)
- Light work with two arms: filing metal, planing wood, raking of a garden
- Moderate work with the body: cleaning a floor, beating a carpet
- Heavy work with the body: railroad track laying, digging, barking trees

Sample Calculation

Assembly line work using a heavy hand tool.

A.	Walking along	2.0 kcal/min
B.	Intermediate value between heavy work with two arms and light work with the body	3.0 kcal/min
	Subtotal:	5.0 kcal/min
C.	Add for basal metabolism	1.0 kcal/min
	Total:	6.0 kcal/min

E. The thermometers must be so placed that the readings are representative of the condition where the men work or rest, respectively.

The methodology outlined above is more fully explained by Minard.[3,4]

II. *Work Load Categories*

Heat produced by the body and the environmental heat together determine the total heat load. Therefore, if work is to be performed under hot environmental conditions, the workload category of each job shall be established and the heat exposure limit pertinent to the workload evaluated against the applicable standard in order to protect the worker exposure beyond the permissible limit.

A. The work load category may be established by ranking each job into light, medium, and heavy categories on the basis of type of operation. Where the work load is ranked into one of said three categories, i.e.,

(1) light work (up to 200 kcal/hr or 800 Btu/hr): e.g., sitting or standing to control machines, performing light hand or arm work,

(2) moderate work (200-350 kcal/hr or 800-1400 Btu/hr): e.g., walking about with moderate lifting and pushing, or

(3) heavy work (350-500 kcal/hr or 1400-2000 Btu/hr): e.g., pick and shovel work,

The permissible heat exposure limit for that workload shall be determined from Table 1.

B. The ranking of the job may be performed either by measuring the worker's metabolic rate while performing his job or by estimating his metabolic rate with the use of Tables 2 and 3. Additional tables available in the literature[5-8] may be utilized also. When this method is used the permissible heat exposure limit can be determined by Figure 1.

III. *Work-Rest Regimen*

The permissible exposure limits specified in Table 1 and Figure 1 are based on the assumption that the WBGT value of the resting place is the same or very close to that of the workplace. Where the WBGT of the work area is different from that of the rest area a time-weighted average value should be used for both environmental and metabolic heat. When time-weighted average values are used, the appropriate curve on Figure 1 is the solid line labeled "continuous."

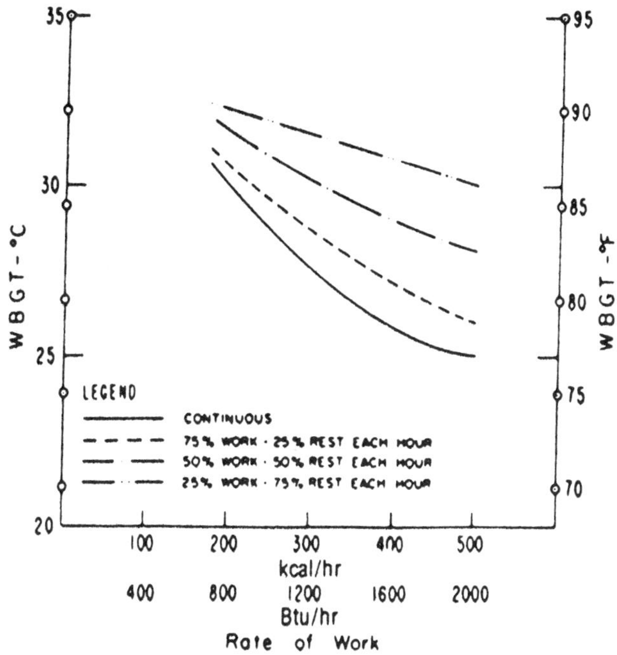

Figure 1—Permissible Heat Exposure Threshold Limit Values.

The time-weighted average metabolic rate (M) shall be determined by the equation:

$$\text{Av. M} = \frac{M_1 \times t_1 + M_2 \times t_2 + \ldots + M_n \times t_n}{t_1 + t_2 + \ldots + t_n}$$

where M_1, M_2...and M_n are estimated or measured metabolic rates for the various activities and rest periods of the worker during the time periods t_1, t_2...and t_n (in minutes) as determined by a time study.

The time-weighted average WBGT shall be determined by the equation:

$$\text{Av. WGBT} = \frac{\text{WBGT}_1 \times t_1 + \text{WBGT}_2 \times t_2 + \ldots + \text{WBGT}_n \times t_n}{t_1 + t_2 + \ldots + t_n}$$

where WBGT_1, WBGT_2 ... and WBGT_n are calculated values of WBGT for the various work and rest areas occupied during total time periods; t_1, t_2 ... and t_n are the elapsed times in minutes spent in the corresponding areas which are determined by a time study. Where exposure to hot environmental conditions is continuous for several hours or the entire work day, the time-weighted averages shall be calculated as hourly time-weighted average, i.e., $t_1 + t_2 + \ldots t_n = 60$ minutes. Where the exposure is intermittent, the time-weighted averages shall be calculated as two-hour time-weighted averages, i.e., $t_1 + t_2 + \ldots + t_n = 120$ minutes.

The permissible exposure limits for continuous work are applicable where there is a work-rest regimen of a 5-day work week and an 8-hour work day with a short morning and afternoon break (approximately 15 minutes) and a longer lunch break (approximately 30 minutes). Higher exposure limits are permitted if additional resting time is allowed. All breaks, including unscheduled pauses and administrative or operational waiting periods during work, may be counted as rest time when additional rest allowance must be given because of high environmental temperatures.

IV. *Water and Salt Supplementation*

During the hot season or when the worker is exposed to artificially generated heat, drinking water shall be made available to the workers in such a way that they are stimulated to frequently drink small amounts, i.e., one cup every 15-20 minutes (about 150 ml or 1/4 pint).

The water shall be kept reasonably cool (10°-15°C or 50.0°-60.0°F) and shall be placed close to the workplace so that the worker can reach it without abandoning the work area.

The workers should be encouraged to salt their food abundantly during the hot season and particularly during hot spells. If the workers are unacclimatized, salted drinking water shall be made available in a concentration of 0.1% (1 g NaCl to 1.0 liter or 1 level tablespoon of salt to 15 quarts of water). The added salt shall be completely dissolved before the water is distributed, and the water shall be kept reasonably cool.

V. *Other Considerations*

A. *Clothing:* The permissible heat exposure TLVs are valid for light summer clothing as customarily worn by workers when working under hot environmental conditions. If special clothing is required for performing a particular job and this clothing is heavier or it impedes sweat evaporation or has higher insulation value, the worker's heat tolerance is reduced, and the permissible heat exposure limits indicated in Table 1 and Figure 1 are not applicable. For each job category where special clothing is required, the permissible heat exposure limit shall be established by an expert.

B. *Acclimatization and Fitness:* Acclimatization to heat involves a series of physiological and psychological adjustments that occur in an individual during this first week of exposure to hot environmental conditions. The recommended heat stress TLVs are valid for acclimated workers who are physically fit. Extra caution must be employed when unacclimated or physically unfit workers must be exposed to heat stress conditions.

References

1. *Health Factors Involved in Working Under Conditions of Heat Stress.* WHO Technical Report Series No. 412 (1969).
2. **Dukes-Dobos, F.N. and A. Henschel:** Development of Permissible Heat Exposure Limits for Occupational Work. *ASHRAE Journal 15(9)*:57-62 (Sept. 1973).
3. **Minard, D.:** *Prevention of Heat Casualities in Marine Corps Recruits, Period of 1955-60, with Comparative Incidence Rates and Climatic Heat Stresses in Other Training Categories.* Research Report No. 4, Contract No. MR 005.01-0001.01, Naval Medical Research Institute, Bethesda, MD (Feb. 21 1961). Published in *Military Medicine 126(44)*:261-272 (April 1961).
4. **Minard, D. and R.L. O'Brien:** *Heat Casualties in the Navy and the Marine Corps 1959-1962 with Appendices on the Field Use of the Wet-Bulb Globe Temperature Index.* Research Report No. 7, Contract No. MR 005.01-0001.01, Naval Medical Research Institute, Bethesda, MD (March 12, 1964).
5. **Astrand, Per-Olof and Kaare Rodahl:** *Textbook of Work Physiology.* McGraw-Hill Book Co., New York, San Francisco (1970).
6. Ergonomics Guide to Assessment of Metabolic and Cardiac Costs of Physical Work. *Am. Ind. Hyg. Assoc. J. 32*:560 (1971).
7. *Energy Requirements for Physical Work.* Research Progress Report No. 30. Purdue Farm Cardiac Project, Agricultural Experiment Station, West Lafayette, IN (1961).
8. **Durnin, J.V.G.A. and R. Passmore:** *Energy, Work and Leisure.* Heinemann Educational Books, Ltd., London (1967).
9. **Lehmann, G.E., A. Muller and H. Spitzer:** Der Kalorienbedarb bie Gewerblicher Arbeit. *Arbeitsphysiol. 14*:166 (1950).

COLD STRESS

These Threshold Limit Values (TLVs) are intended to protect workers from the severest effects of cold stress (hypothermia) and cold injury and to describe exposures to cold working conditions under which it is believed that nearly all workers can be repeatedly exposed without adverse health effects. The TLV objective is to prevent the deep body core temperature from falling below 36°C and to prevent cold injury to body extremities. Deep body temperature is the core temperature of the body as determined by rectal temperature measurements. For a single, occasional exposure to a cold environment, a drop in core temperature to no lower that 35°C should be permitted. In addition to provisions for total body protection, the TLV objective is to protect all parts of the body with emphasis on hands, feet, and head from cold injury.

Introduction

Fatal exposures to cold among workers have almost always resulted from accidental exposures involving failure to escape from low environmental air temperatures or from immersion in low temperature water. The single most important aspect of life-threatening hypothermia is the fall in the deep core temperature of the body. The clinical presentations of victims of hypothermia are shown in Table 4 (taken from Dembert in *AFP,* January 1982). Workmen should be protected from exposure to cold so that the deep core temperature does not fall below 36°C (96.8°F); lower body temperatures will very likely result in reduced mental alertness, reduction in rational decision making, or loss of consciousness with the threat of fatal consequences.

Pain in the extremities may be the first early warning of danger to cold stress. During exposure to cold, maximum severe shivering develops when the body temperature has fallen to 35°C (95°F). This must be taken as a sign of danger to the workers and exposure to cold should be immediately terminated for any workers when severe shivering becomes evident. Useful physical or mental work is limited when severe shivering occurs.

Since prolonged exposure to cold air, or to immersion in cold water, at temperatures well above freezing can lead to dangerous hypothermia, whole body protection must be provided.

1. Adequate insulating clothing to maintain core temperatures above 36°C must be provided to workers if work is performed in air temperatures below 4°C (40°F). Wind chill factor[A] or

[A] Wind chill factor is a unit of heat loss from a body defined in watts per meter squared per hour being a function of the air temperature and wind velocity upon the exposed body.

TABLE 4
Progressive Clinical Presentations of Hypothermia*

Core Temperature °C	Core Temperature °F	Clinical Signs
37.6	99.6	"Normal" rectal temperature
37	98.6	"Normal" oral temperature

TABLE 4, continued

Core Temperature °C	Core Temperature °F	Clinical Signs
36	96.8	Metabolic rate increases in an attempt to compensate for heat loss
35	95.0	Maximum shivering
34	93.2	Victim conscious and responsive, with normal blood pressure
33	91.4	Severe hypothermia below this temperature
32 } 31	89.6 } 87.8	Consciousness clouded; blood pressure becomes difficult to obtain; pupils dilated but react to light; shivering ceases
30 } 29	86.0 } 84.2	Progressive loss of consciousness; muscular rigidity increases; pulse and blood pressure difficult to obtain; respiratory rate decreases
28	82.4	Ventricular fibrillation possible with myocardial irritability
27	80.6	Voluntary motion ceases; pupils nonreactive to light; deep tendon and superficial reflexes absent
26	78.8	Victim seldom conscious
25	77.0	Ventricular fibrillation may occur spontaneously
24	75.2	Pulmonary edema
22 } 21	71.6 } 69.8	Maximum risk of ventricular fibrilation
20	68.0	Cardiac standstill
18	64.4	Lowest accidental hypothermia victim to recover
17	62.6	Isoelectric electroencephalogram
9	48.2	Lowest artificially cooled hypothermia patient to recover

* Presentations approximately related to core temperature. Reprinted from the January 1982 issue of *American Family Physician*, published by the American Academy of Family Physicians.

the cooling power of the air is a critical factor. The higher the wind speed and the lower the temperature in the work area, the greater the insulation value of the protective clothing required. An equivalent chill temperature chart relating the actual dry bulb air temperature and the wind velocity is presented in Table 5. The equivalent chill temperature should be used when estimating the combined cooling effect of wind and low air temperatures on exposed skin or when determining clothing insulation requirements to maintain the deep body core temperature.

2. Unless there are unusual or extenuating circumstances, cold injury to other than hands, feet, and head is not likely to occur without the development of the initial signs of hypothermia. Older workers or workers with circulatory problems require special precautionary protection against cold injury. The use of extra insulating clothing and/or a reduction in the duration of the exposure period are among the special precautions which should be considered. The precautionary actions to be taken will depend upon the physical condition of the worker and should be determined with the advice of a physician with knowledge of the cold stress factors and the medical condition of the worker.

Evaluation and Control

For exposed skin, continuous exposure should not be permitted when the air speed and temperature results in an equivalent chill temperature of –32°C (–25°F). Superficial or deep local tissue freezing will occur only at temperatures below –1 °C regardless of wind speed.

At air temperatures of 2°C (35.6°F) or less, it is imperative that workers who become immersed in water or whose clothing becomes wet be immediately provided a change of clothing and be treated for hypothermia.

Recommended limits for properly clothed workers for periods of work at temperatures below freezing are shown in Table 6.

Special protection of the hands is required to maintain manual dexterity for the prevention of accidents:

1. If fine work is to be performed with bare hands for more than 10-20 minutes in an environment below 16°C (60°F), special provisions should be established for keeping the workers' hands warm. For this purpose, warm air jets, radiant heaters (fuel burner or electric radiator), or contact warm plates may be utilized. Metal handles of tools and control bars shall be covered by thermal insulating material at temperatures below –1°C (30°F).

TABLE 5

Cooling Power of Wind on Exposed Flesh Expressed as Equivalent Temperature (under calm conditions)*

Estimated Wind Speed (in mph)	Actual Temperature Reading (°F)											
	50	40	30	20	10	0	-10	-20	-30	-40	-50	-60
	Equivalent Chill Temperature (°F)											
calm	50	40	30	20	10	0	-10	-20	-30	-40	-50	-60
5	48	37	27	16	6	-5	-15	-26	-36	-47	-57	-68
10	40	28	16	4	-9	-24	-33	-46	-58	-70	-83	-95
15	36	22	9	-5	-18	-32	-45	-58	-72	-85	-99	-112
20	32	18	4	-10	-25	-39	-53	-67	-82	-96	-110	-121
25	30	16	0	-15	-29	-44	-59	-74	-88	-104	-118	-133
30	28	13	-2	-18	-33	-48	-63	-79	-94	-109	-125	-140
35	27	11	-4	-20	-35	-51	-67	-82	-98	-113	-129	-145
40	26	10	-6	-21	-37	-53	-69	-85	-100	-116	-132	-148
(Wind speeds greater than 40 mph have little additional effect.)	*LITTLE DANGER* In < hr with dry skin. Maximum danger of false sense of security				*INCREASING DANGER* Danger from freezing of exposed flesh within one minute.				*GREAT DANGER* Flesh may freeze within 30 seconds.			
	Trenchfoot and immersion foot may occur at any point on this chart.											

* Developed by U.S. Army Research Institute of Environmental Medicine, Natick, MA.

2. If the air temperature falls below 16°C (60°F) for sedentary, 4°C (40°F) for light, –7°C (20°F) for moderate work and fine manual dexterity is not required, then gloves shall be used by the workers.

To prevent contact frostbite, the workers should wear anti-contact gloves.

1. When cold surfaces below –7°C (20°F) are within reach, a warning should be given to each worker by his supervisor to prevent inadvertent contact by bare skin.
2. If the air temperature is –17.5°C (0°F) or less, the hands should be protected by mittens. Machine controls and tools for use in cold conditions should be designed so that they can be handled without removing the mittens.

Provisions for additional total body protection are required if work is performed in an environment at or below 4°C (40°F). The workers shall wear cold protective clothing appropriate for the level of cold and physical activity:

1. If the air velocity at the job site is increased by wind, draft, or artificial ventilating equipment, the cooling effect of the wind shall be reduced by shielding the work area, or by wearing an easily removable outer windbreak layer garment. Wind chill cooling rates are illustrated in Figure 2 and Table 7.
2. If only light work is involved and if the clothing on the worker may become wet on the job site, the outer layer of the clothing in use may be of a type impermeable to water. With more severe work under such conditions, the outer layer should be water repellent, and the outerwear should be changed as it becomes wetted. The outer garments must include provisions for easy ventilation in order to prevent wetting of inner layers by sweat. If work is done at normal temperatures or in a hot environment before entering the cold area, the employee shall make sure that his clothing is not wet as a consequence of sweating. If his clothing is wet, the employee shall change into dry clothes before entering the cold area. The workers shall change socks and any removable felt insoles at regular daily intervals or use vapor barrier boots. The optimal frequency of change shall be determined empirically and will vary individually and according to the type of shoe worn and how much the individual's feet sweat.
3. If extremities, ears, toes, and nose, cannot be protected sufficiently to prevent sensation of excessive cold or frostbite by handware, footwear, and face masks, these protective items shall be supplied in auxiliary heated versions.
4. If the available clothing does not give adequate protection to prevent hypothermia or frostbite, work shall be modified or

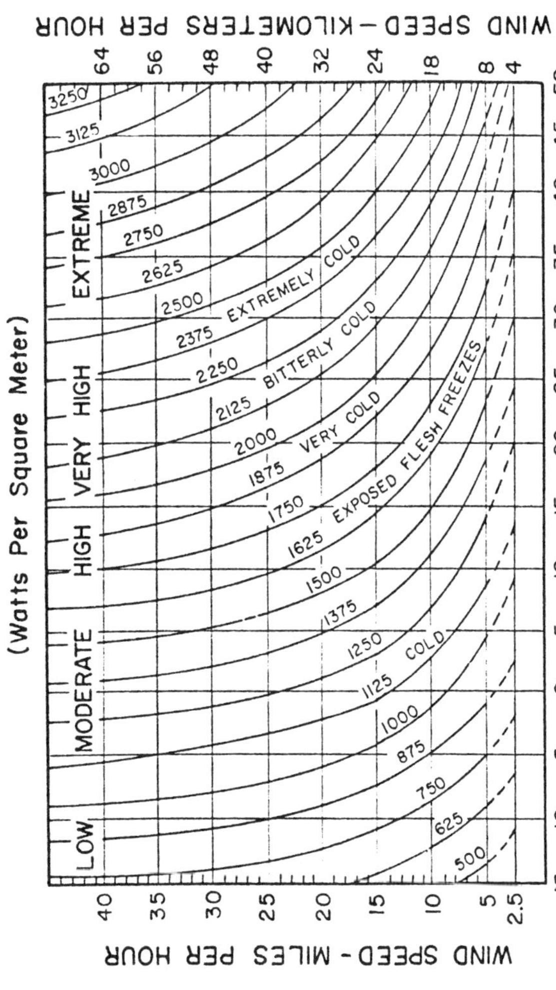

Figure 2—Wind chill cooling rates. Adapted from Canadian Department of the Environment, Atmospheric Environment Service.

TABLE 6
Threshold Limit Values Work/Warm-up Schedule for Four-Hour Shift*

Air Temperature—Sunny Sky		No Noticeable Wind		5 mph Wind		10 mph Wind		15 mph Wind		20 mph Wind	
°C (approx.)	°F	Max. Work Period	No. of Breaks	Max. Work Period	No. of Breaks	Max. Work Period	No. of Breaks	Max. Work Period	No. of Breaks	Max. Work Period	No. of Breaks
1. -26° to -28°	-15° to -19°	(Norm. Breaks)	1	(Norm. Breaks)	1	75 min	2	55 min	3	40 min	4
2. -29° to -31°	-20° to -24°	(Norm. Breaks)	1	75min	2	55 min	3	40 min	4	30 min	5
3. -32° to -34°	-25° to -29°	75 min	2	55 min	3	40 min	4	30 min	5	Non-emergency work should cease	
4. -35° to -37°	-30° to -34°	55 min	3	40 min	4	30 min	5	Non-emergency work should cease			
5. -38° to -39°	-35° to -39°	40 min	4	30 min	5	Non-emergency work should cease					
6. -40° to -42°	-40° to -44°	30 min	5	Non-emergency work should cease							
7. -43° & below	-45° & below	Non-emergency work should cease		↓		↓		↓		↓	

Notes for Table 6:

1. Schedule applies to moderate to heavy work activity with warm-up breaks of ten (10) minutes in a warm location. For Light-to-Moderate Work (limited physical movement): apply the schedule one step lower. For example, at −30°F with no noticeable wind (Step 4), a worker at a job with little physical movement should have a maximum work period of 40 minutes with 4 breaks in a 4-hour period (Step 5).
2. The following is suggested as a guide for estimating wind velocity if accurate information is not available:
   5 mph: light flag moves; 10 mph: light flag fully extended; 15 mph: raises newspaper sheet; 20 mph: blowing and drifting snow.
3. If only the wind chill cooling rate is available, a rough rule of thumb for applying it rather than the temperature and wind velocity factors given above would be: 1) special warm-up breaks should be initiated at a wind chill of about 1750 W/m^2; 2) all non-emergency work should have ceased at or before a wind chill of 2250 W/m^2. In general the warm-up schedule provided above slightly under-compensates for the wind at the warmer temperatures, assuming acclimatization and clothing appropriate for winter work. On the other hand, the chart slightly over-compensates for the actual temperatures in the colder ranges, since windy conditions rarely prevail at extremely low temperatures.

* Adapted from Occupational Health & Safety Division, Saskatchewan Department of Labour.

TABLE 7
Wind Chill Cooling Rate Effects*

Wind Chill Rates (Watts/m²)	Comments/Effects
700	Conditions considered comfortable when dressed for skiing.
1200	Conditions no longer pleasant for outdoor activities on overcast days.
1400	Conditions no longer pleasant for outdoor activities on sunny days.
1600	Freezing of exposed skin begins for most people depending on the degree of activity and the amount of sunshine.
2300	Conditions for outdoor travel such as walking become dangerous. Exposed areas of the face freeze in less than 1 minute for the average person.
2700	Exposed flesh will freeze within half a minute for the average person.

* Adapted from Canadian Department of the Environment, Atmospheric Environment Service.

suspended until adequate clothing is made available or until weather conditions improve.

5. Workers handling evaporative liquid (gasoline, alcohol or cleaning fluids) at air temperatures below 4°C (40°F) shall take special precautions to avoid soaking of clothing or gloves with the liquids because of the added danger of cold injury due to evaporative cooling. Special note should be taken of the particularly acute effects of splashes of "cryogenic fluids" or those liquids with a boiling point only just above ambient temperatures.

Work–Warming Regimen

If work is performed continuously in the cold at an equivalent chill temperature (ECT) or below −7°C (20°F) heated warming shelters (tents, cabins, rest rooms, etc.) shall be made available nearby and the workers should be encouraged to use these shelters at regular intervals, the frequency depending on the severity of the environmental exposure. The onset of heavy shivering, frostnip, the feeling of excessive fatigue, drowsiness, irritability, or euphoria are indications for immediate return to the shelter. When entering the heated shelter, the outerlayer of clothing shall be removed and the remainder of the clothing loosened to permit sweat evaporation or a change of dry work clothing provided. A change of dry work clothing shall be provided as necessary to prevent workers from returning to their work with wet clothing. Dehydration, or the loss of body fluids, occurs insidiously in the cold environment and may increase the susceptibility of the worker to cold injury due to a significant change in blood flow to the extremities. Warm sweet drinks and soups should be provided at the work site to provide caloric intake and fluid volume. The intake of coffee should be limited because of the diuretic and circulatory effects.

For work practices at or below −12°C (10°F) ECT, the following shall apply:

1. The worker shall be under constant protective observation (buddy system or supervision).
2. The work rate should not be so high as to cause heavy sweating that will result in wet clothing; if heavy work must be done, rest periods must be taken in heated shelters and opportunity for changing into dry clothing shall be provided.
3. New employees shall not be required to work fulltime in cold in the first days until they become accustomed to the working conditions and required protective clothing.
4. The weight and bulkiness of clothing shall be included in estimating the required work performance and weights to be lifted by the worker.
5. The work shall be arranged in such a way that sitting still or standing still for long periods is minimized. Unprotected metal chair seats shall not be used. The worker should be protected from drafts to the greatest extent possible.
6. The workers shall be instructed in safety and health procedures. The training program shall include as a minimum instruction in:
   a. Proper rewarming procedures and appropriate first aid treatment.
   b. Proper clothing practices.
   c. Proper eating and drinking habits.
   d. Recognition of impending frostbite.
   e. Recognition of signs and symptoms of impending hypothermia or excessive cooling of the body even when shivering does not occur.
   f. Safe work practices.

Special Workplace Recommendations

Special design requirements for refrigerator rooms include the following:

1. In refrigerator rooms, the air velocity should be minimized as much as possible and should not exceed 1 meter/sec (200 fpm) at the job site. This can be achieved by properly designed air distribution systems.
2. Special wind protective clothing shall be provided based upon existing air velocities to which workers are exposed.

Special caution shall be exercised when working with toxic substances and when workers are exposed to vibration. Cold exposure may require reduced exposure limits.

Eye protection for workers employed out-of-doors in a snow and/or ice-covered terrain shall be supplied. Special safety goggles to protect against ultraviolet light and glare (which can produce temporary conjunctivitis and/or temporary loss of vision) and blowing ice crystals are required when there is an expanse of snow coverage causing a potential eye exposure hazard.

Workplace monitoring is required as follows:

1. Suitable thermometry should be arranged at any workplace where the environmental temperature is below 16°C (60°F) to enable overall compliance with the requirements of the TLV to be maintained.
2. Whenever the air temperature at a workplace falls below −1°C (30°F), the dry bulb temperature should be measured and recorded at least every 4 hours.
3. In indoor workplaces, the wind speed should also be recorded at least every 4 hours whenever the rate of air movement exceeds 2 meters per second (5 mph).
4. In outdoor work situations, the windspeed should be measured and recorded together with the air temperature whenever the air temperature is below −1°C (30°F).
5. The equivalent chill temperature shall be obtained from Table 16 in all cases where air movement measurements are required, and shall be recorded with the other data whenever the equivalent chill temperature is below −7°C (20°F).

Employees shall be excluded from work in cold at −1°C (30°F) or below if they are suffering from diseases or taking medication which interferes with normal body temperature regulation or reduces tolerance to work in cold environments. Workers who are routinely exposed to temperatures below −24°C (−10°F) with wind speeds less than five miles per hour, or air temperatures below −18°C (0°F) with wind speeds above five miles per hour, should be medically certified as suitable for such exposures.

Trauma sustained in freezing or subzero conditions requires special attention because an injured worker is predisposed to secondary cold injury. Special provisions must be made to prevent hypothermia and secondary freezing of damaged tissues in addition to providing for first aid treatment.

HAND-ARM (SEGMENTAL) VIBRATION

These Threshold Limit Values (TLVs) in Table 8 refer to component accelerations levels and durations of exposure that represent conditions under which it is believed that most workers may be exposed repeatedly without progressing to Stage 3 of the Taylor-Pelmear Classification System for Vibration-induced White Finger (VWF), also known as Raynaud's Phenomenon of Occupational Origin. Since there is a paucity of dose–response relationships for VWF, these recommendations have been derived from epidemiological data from forestry, mining, and metal working. These values should be used as guides in the control of hand–arm vibration exposure and because of individual susceptibility, should not be regarded as defining a boundary between safe and dangerous levels.

It should be recognized that the application of the TLV alone for hand–arm vibration will not protect all workers from the adverse effects of hand–arm vibration exposure. The use of: 1) antivibration tools, 2) antivibration gloves, 3) proper work practices which keep the worker's hands and remaining body warm and also minimize the vibration coupling between the worker and the vibration tool are necessary to minimize vibration exposure, and 4) a conscienciously applied medical surveillance program are ALL necessary to rid VWF from the workplace.

Continuous, Intermittent, Impulsive, or Impact Hand–Arm Vibration

The measurement of vibration should be performed in accordance with the procedures and instrumentation specified by the International Standard ISO 5349 (1986), *Guide for the Measurement and the Assessment of Human Exposure to Vibration Transmitted to the Hand,* and summarized below:

The acceleration of a vibration handle or work piece should be determined in three mutually orthogonal directions at a point close to where vibration enters the hand. The directions shall preferably be those forming the ISO biodynamic coordinate system but may be a closely related basicentric system with its origin at the interface between the hand and the vibrating surface (see Figure 3) to accommodate different handle or work piece configurations. A small and lightweight transducer shall be mounted so as to record

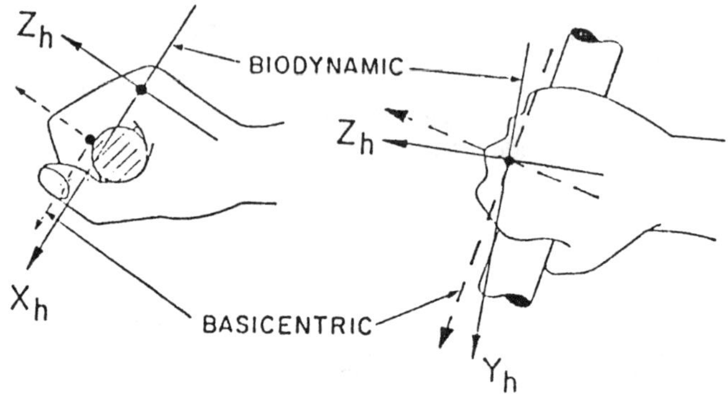

Figure 3—Biodynamic and basicentric coordinate systems for the hand, showing the directions of the acceleration components (ISO 5349).

accurately one or more orthogonal components of the source vibration in the frequency range from 5 to 1500 Hz. Each component should be frequency-weighted by a filter network with gain characteristics specified by the ISO for human-response vibration measuring instrumentation, to account for the change in vibration hazard with frequency (see Figure 4).

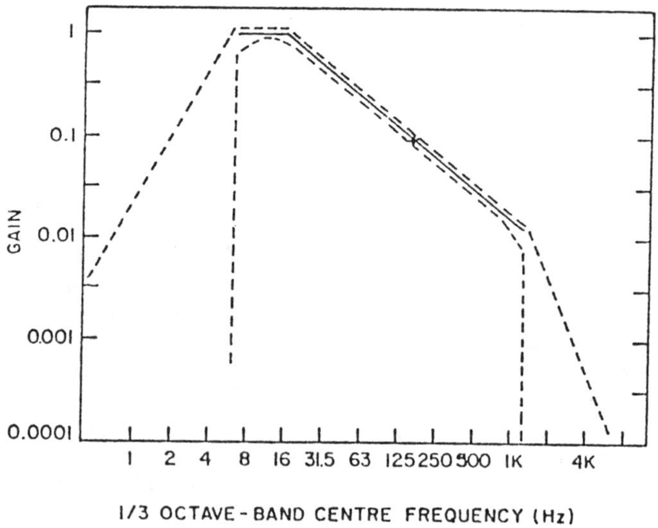

Figure 4—Gain characteristics on the filter network used to frequency-weight acceleration components (continuous line). The filter tolerances (dashed lines) are provisional and are those contained in ISO 5349.

Assessment of vibration exposure should be made for EACH applicable direction (X_h, Y_h, Z_h) since vibration is a vector quantity (magnitude and direction). In each direction, the magnitude of the vibration during normal operation of the power tool, machine or work piece shall be expressed by the root-mean-square (rms) value of the frequency-weighted component accelerations, in units of meters per second squared (m/s^2), or gravitational units (g), the largest of which, a_K, forms the basis for exposure assessment.

For each direction being measured, linear integration shall be employed for vibrations that are of extremely short duration or vary substantially in time. If the total daily vibration exposure in a given direction is composed of several exposures at different rms accelerations, then the equivalent, frequency-weighted component acceleration in that direction shall be determined in accordance with the following equation:

$$\left(a_{K_{eq}}\right) = \left[\frac{1}{T}\sum_{i=1}^{n}\left(a_{K_i}\right)^2 T_i\right]^{1/2}$$

$$= \sqrt{\left(a_{K_1}\right)^2 \frac{T_1}{T} + \left(a_{K_2}\right)^2 \frac{T_2}{T} + \ldots \left(a_{K_n}\right)^2 \frac{T_n}{T}}$$

where: $T = \sum_{i=1}^{n} T_i$

T = total daily exposure duration

a_{K_i} = ith frequency-weighted, rms acceleration component with duration T_i

These computations may be performed by commercially available human-response vibration measuring instruments.

TABLE 8
Threshold Limit Values for Exposure of the Hand to Vibration in Either X_h, Y_h, Z_h Directions

Total Daily Exposure Duration[a]	Values of the Dominant,[b] Frequency-Weighted, rms, Component Acceleration Which Shall not be Exceeded a_K, ($a_{K_{eq}}$)	
	m/s^2	g[c]
4 hours and less than 8	4	0.40
2 hours and less than 4	6	0.61
1 hour and less than 2	8	0.81
less than 1 hour	12	1.22

[a] The total time vibration enters the hand per day, whether continuously or intermittently.
[b] Usually one axis of vibration is dominant over the remaining two axis. If one or more vibration axis exceeds the Total Daily Exposure then the TLV has been exceeded.
[c] $g = 9.81\ m/s^2$.

Notes: Table 8:

1. The weighting network provided in Figure 4 is considered the best available to frequency weight acceleration components.[1-9] However, recent studies suggest that the frequency weighting at higher frequencies (above 16 Hz) may not incorporate a sufficient safety factor, and CAUTION must be applied when tools with high-frequency components are used.
2. Acute exposures to frequency-weighted, rms, component accelerations in excess of the TLVs for infrequent periods of time (e.g., 1 day per week, or several days over a 2-week period) are not necessarily more harmful.[2-4]
3. Acute exposures to frequency-weighted, rms, component accelerations of three times the magnitude of the TLVs are expected to result in the same health effects after between 5 and 6 years of exposure.[2-4]
4. Preventive measures, including specialized preemployment and annual medical examinations to identify persons susceptible to vibration, should be implemented in situations in which workers are or will be exposed to hand-arm vibration.[3-7]

5. To moderate the adverse effects of vibration exposure, workers should be advised to avoid *continuous* vibration exposure by cessation of vibration exposure for approximately 10 minutes per continuous vibration hour.
6. Good work practices should be used and should include instructing workers to employ a minimum hand grip force consistent with safe operation of the power tool or process, keep their body and hands warm and dry, and avoid smoking.[2,3]
7. A transducer and its device for attachment to the vibrating source suitable for measurement purposes together should weigh less than 15 grams and should possess a cross-axis sensitivity of less than 10%.
8. The measurement by many (mechanically underdamped) piezoelectric accelerometers of repetitive, large displacement, impulsive vibrations, such as those produced by percussive pneumatic tools, is subject to error. The insertion of a suitable, low-pass, mechanical filter between the accelerometer and the source of vibration with a cut-off frequency of 1500 Hz or greater (and cross-axis sensitivity of less than 10%) can help eliminate incorrect readings.[3,4]
9. The manufacturer and type number of all apparatus used to measure vibration should be reported, as well as the value of the dominant direction and frequency-weighted, rms, component acceleration.

References

1. **Pyykko I.:** Vibration Syndrome. A Review. In: *Vibration and Work,* pp. 1-24. O. Dorhonen, Ed. Institute of Occupational Health, Helsinki (1976).
2. *Vibration White Finger in Industry,* W. Taylor and P.L. Pelmear, Eds. Academic Press, London (1975).
3. **NIOSH:** *Proceedings of the International Occupational Hand-Arm Vibration Conference,* D.E. Wasserman and W. Taylor, Eds. DHEW (NIOSH) Pub. No. 77-170 (1977).
4. **Brammer, A.J.:** Threshold Limit for Hand-Arm Vibration Exposure Throughout the Workday. In: *Vibration Effects on the Hand and Arm in Industry,* pp. 291-301. A.J. Brammer and W. Taylor, Eds. John Wiley & Sons, New York (1982).
5. **Wasserman, D.E. and W. Taylor:** Environmental and Occupational Medicine. In: *Occupational Vibration,* Chap. 68, pp. 743-749. W.N. Rom, Ed. Little, Brown and Co., Boston (1982).
6. **NIOSH:** *Current Intelligence Bulletin #38: Vibration Syndrome.* DHHS (NIOSH) Pub. No. 83-110 (1983).
7. **NIOSH:** *Vibration Syndrome.* NIOSH Videotape #177 (27 minutes). Cincinnati, OH.
8. **International Organization for Standardization:** *Guide for the Measurement and the Assessment of Human Exposure to Vibration Transmitted to the Hand.* ISO 5349. International Organization for Standardization, Geneva (1986).
9. **International Organization for Standardization:** *Human-Response Vibration Measuring Instrumentation.* Second Draft Proposal DP 8041. ISO/TC 108/SC 3 n 99. International Organization for Standardization, Geneva (unpublished, 1982).

IONIZING RADIATION

The Physical Agents TLV Committee accepts the philosophy and recommendations of the National Council on Radiation Protection and Measurements (NCRP) for the ionizing radiation TLV. The NCRP is charted by Congress to, in part, collect, analyze, develop, and disseminate information and recommendations about protection against radiation and about radiation measurements, quantities, and units, including development of basic concepts in these areas. NCRP Report No. 39 provides basic philosophy and concepts leading to protection criteria established in the same report.[1] Other NCRP reports address specific areas of radiation protection and, collectively, provide an excellent basis for establishing a sound program for radiation control. The Committee recommends the listed references as substantive documentation of a sound basis for ionizing radiation protection. The Committee also strongly recommends that all exposure to ionizing radiation be kept as low as reasonably achievable.

References

1. *Basic Radiation Protection Criteria.* NCRP Report No. 39 (January 15, 1971).
2. *Maximum Permissible Body Burdens and Maximum Permissible Concentrations of Radionuclides in Air and in Water for Occupational Exposure.* National Bureau of Standards Handbook 69, (June 5, 1959), with *Addendum 1* (August 1963). Available as NCRP Report No. 22.

The above documents, as well as information on numerous other NCRP Reports addressing specific subjects in ionizing radiation protection, are available from: NCRP Publications, 7910 Woodmont Ave., Suite 1016, Bethesda, MD 20814.

LASERS

The Threshold Limit Values (TLVs) are for exposure to laser radiation under conditions to which nearly all workers may be exposed without adverse effects. The values should be used as guides in the control of exposures and should not be regarded as fine lines between safe and dangerous levels. They are based on the best available information from experimental studies.

Limiting Apertures

The TLVs expressed as radiant exposure or irradiance in this section may be averaged over an aperture of 1 mm except for TLVs for the eye in the spectral range of 400–1400 nm, which should be averaged over a 7-mm limiting aperture (pupil), and except for all TLVs for wavelengths between 0.1–1 mm where the limiting aperture is 10 mm. No modification of the TLVs is permitted for pupil sizes less than 7 mm.

The TLVs for "extended sources" apply to sources which subtend an angle greater than α (Table 9) which varies with exposure time. This angle is *not* the beam divergence of the source.

TABLE 9
Limiting Angle to Extended Source (400–1400 nm) Which May Be Used For Applying Extended Source TLVs

Exposure Duration(s)	Angle α (mrad)	Exposure Duration(s)	Angle α (mrad)
10^{-9}	8.0*	10^{-2}	5.7
10^{-8}	5.4*	10^{-1}	9.2
10^{-7}	3.7*	1.0	15
10^{-6}	2.5*	10	24
10^{-5}	1.7*	10^{2}	24
10^{-4}	2.2	10^{3}	24
10^{-3}	3.6	10^{4}	24

* Values for durations less than 10^{-4} s at wavelengths from 1050–1400 nm should be increased by a fctor of 1.4.

Correction Factors A and B (C_A and C_B)

The TLVs for ocular exposure in Tables 10 and 11 are to be used as given for all wavelength ranges. The TLVs for wavelengths between 700 nm and 1049 nm are to be increased by a uniformly extrapolated factor (C_A) as shown in Figure 5. Between 1049 nm and 1400 nm, the TLV has been increased by a factor (C_A) of five. For certain exposure times at wavelengths between 550 nm and 700 nm, a correction factor (C_B) must be applied.

The TLVs for skin exposure are given in Table 12. The TLVs are to be increased by a factor (C_A) as shown in Figure 5 for wavelengths between 700 nm and 1400 nm. To aid in the determination of TLVs for exposure durations requiring calculations of fractional powers Figures 6, 7, and 8 may be used.

Repetitively Pulsed Exposures

Scanned CW lasers or repetitively pulsed lasers can both produce repetitively pulsed exposure conditions. The TLV for

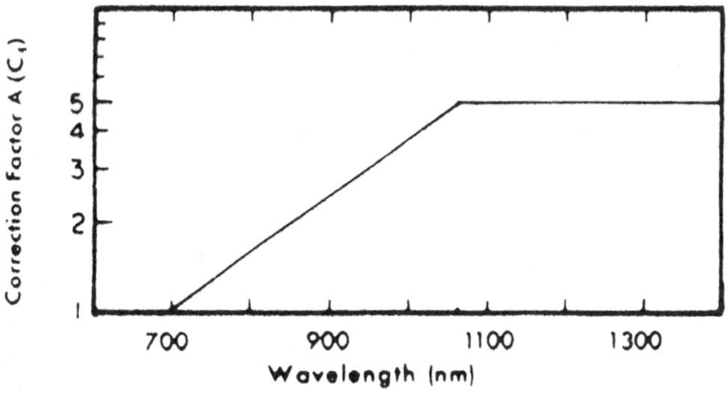

Figure 5—TLV correction factor for λ = 700–1400 nm.* (* For λ = 700–1049 nm, $C_A = 10^{[0.002(\lambda-700)]}$; For λ = 1050–1400 nm, C_A= 5.)

intrabeam viewing which is applicable to wavelengths between 400 and 1400 nm and a single-pulse exposure (of pulse duration *t*) is modified in this instance by a correction factor determined by the number of pulses in the exposure. First, calculate the number of pulses (n) in an expected exposure situation; this is the pulse repetitition frequency (PRF in Hz) multiplied by the duration of exposure. Normally, realistic exposures may range from 0.25 second (s) for a bright visible source to 10 s for an infrared source. The corrected TLV on a per-pulse basis is:

$$\text{TLV} = (n^{-1/4})\ (\text{TLV for single-pulse}) \qquad (1)$$

This approach applies only to thermal-injury conditions, i.e., all exposures at wavelengths greater than 700 nm, and for many exposures at shorter wavelengths. For wavelengths less than or equal to 700 nm, the corrected TLV from equation 1 above applies if the average irradiance does not exceed the TLV for continuous exposure. The average irradiance (i.e., the total accumulated exposure for *nt* s) shall not exceed the radiant exposure given in Table 10 for exposure durations of 10 s to T_1.

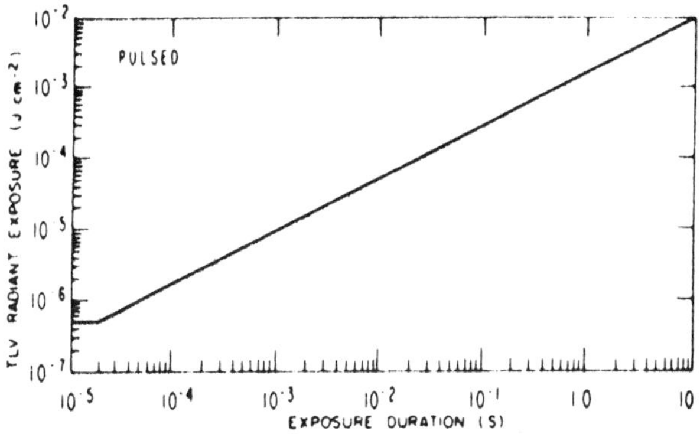

Figure 6a — TLV for intrabeam (direct) viewing of laser beam (400–700 nm).

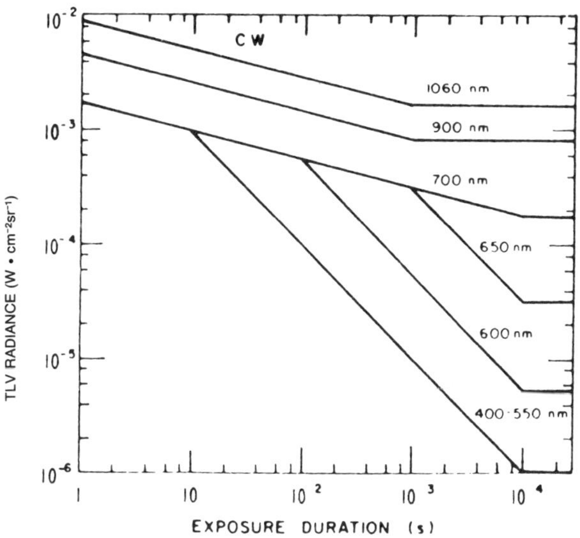

Figure 6b — TLV for intrabeam (direct) viewing of CW laser beam (400–1400 nm).

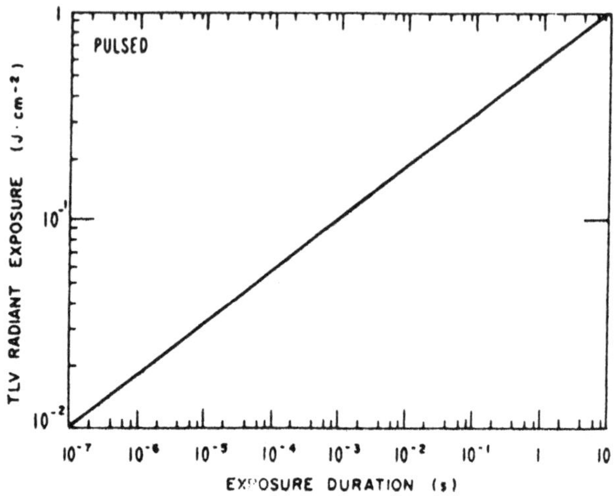

Figure 7a—TLV for laser exposure of skin and eyes for far-infrared radiation (wave-lengths greater than 1.4 μm).

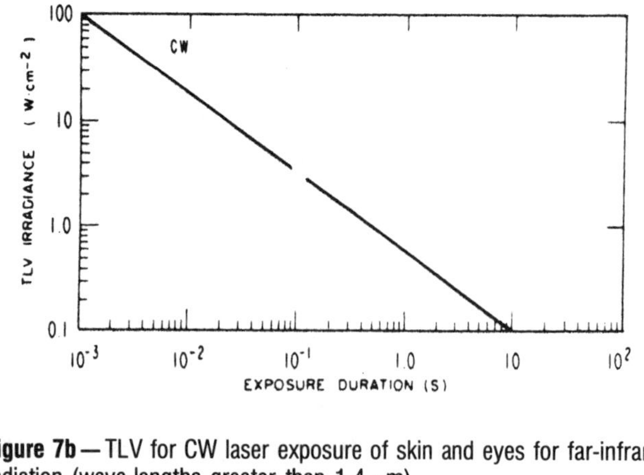

Figure 7b—TLV for CW laser exposure of skin and eyes for far-infrared radiation (wave-lengths greater than 1.4 μm).

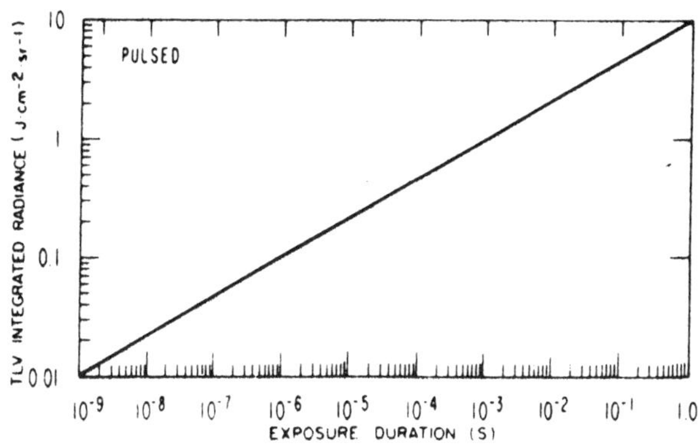

Figure 8a — TLV for extended sources or diffuse reflections of laser radiation (400–700 nm).

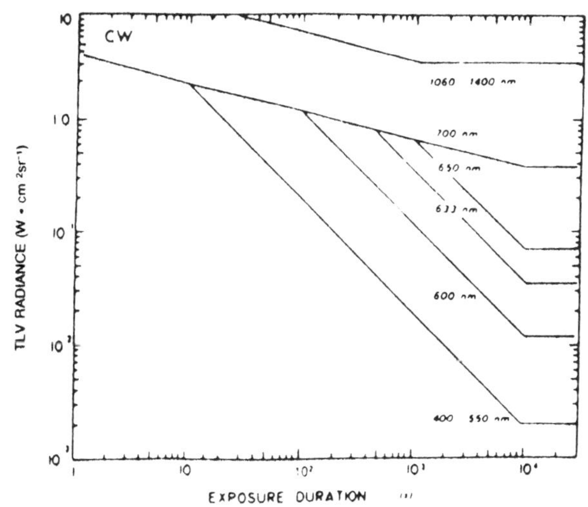

Figure 8b — TLV for extended sources or diffuse reflections of laser radiations (400–1400 nm).

TABLE 10
Threshold Limit Value for Direct Ocular Exposures (Intrabeam Viewing) from a Laser Beam

Spectral Region	Wave Length	Exposure Time, (t) Seconds	TLV		
*UVC	180 nm to 280 nm	10^{-9} to 3×10^{4}	3	$mJ \bullet cm^{-2}$	
UVB	280 nm to 302 nm	"	3	"	not to exceed $0.56\ t^{1/4}\ J \bullet cm^{-2}$ for $t \leqq 10$ s.
	303 nm	"	4	"	
	304 nm	"	6	"	
	305 nm	"	10	"	
	306 nm	"	16	"	
	307 nm	"	25	"	
	308 nm	"	40	"	
	309 nm	"	63	"	
	310 nm	"	100	"	
	311 nm	"	160	"	
	312 nm	"	250	"	
	313 nm	"	400	"	
	314 nm	"	630	"	

UVA	315 nm to 400 nm	10^{-9} to 10	$.56\,t^{1/4}$ J • cm^{-2}
	" "	10 to 10^3	1.0 J • cm^{-2}
	" "	10^3 to 3×10^4	1.0 mW • cm^{-2}
Light	400 nm to 700 nm	10^{-9} to 1.8×10^{-5}	5×10^{-7}J • cm^{-2}
	400 nm to 700 nm	1.8×10^{-5} to 10	$1.8\,(t/\sqrt[4]{t})$ mJ • cm^{-2}
	400 nm to 549 nm	10 to 10^4	10 mJ • cm^{-2}
	550 nm to 700 nm	10 to T_1	$1.8\,(t/\sqrt[4]{t})$ mJ • cm^{-2}
	550 nm to 700 nm	T_1 to 10^4	$10\,C_B$ mJ • cm^{-2}
	400 nm to 700 nm	10^4 to 3×10^4	C_B μW • cm^{-2}
IR-A	700 nm to 1049 nm	10^{-9} to 1.8×10^{-5}	$5\,C_A \times 10^{-7}$ J • cm^{-2}
	700 nm to 1049 nm	1.8×10^{-5} to 10^3	$1.8\,C_A\,(t/\sqrt[4]{t})$ mJ • cm^{-2}
	1050 nm to 1400 nm	10^{-9} to 10^{-4}	5×10^{-6} J • cm^{-2}
	1050 nm to 1400 nm	10^{-4} to 10^3	$9(t/\sqrt[4]{t})$ mJ • cm^{-2}
	700 nm to 1400 nm	10^3 to 3×10^4	$320\,C_A$ μW • cm^{-2}
IR-B & C	1.4 μm to 10^3 μm	10^{-9} to 10^{-7}	10^{-2} J • cm^{-2}
	" "	10^{-7} to 10	$0.56\,\sqrt[4]{t}$ J • cm^{-2}
	" "	10 to 3×10^4	0.1 W • cm^{-2}

C_A – See Fig. 5; $C_B = 1$ for $\lambda = 400$ to 549 nm; $C_B = 10^{[0.015(\lambda - 550)]}$ for $\lambda = 550$ to 700 nm; $T_1 = 10$ s for $\lambda = 400$ to 549 nm; $T_1 = 10 \times 10^{[0.02(\lambda - 550)]}$ for $\lambda = 550$ to 700 n.

TABLE 11
Threshold Limit Values for Viewing a Diffuse Reflection of a Laser Beam or an Extended Source Laser

Spectral Region	Wave Length	Exposure Time, (t) Seconds	TLV
UV	180 nm to 400 nm	10^{-9} to 3×10^{4}	Same as Table 10
Light	400 nm to 700 nm	10^{-9} to 10	$10 \sqrt[3]{t}$ J • cm^{-2} • sr^{-1}
	400 nm to 549 nm	10 to 10^{4}	21 J • cm^{-2} • sr^{-1}
	550 nm to 700 nm	10 to T_1	$3.83 (t/\sqrt[3]{t})$ J • cm^{-2} • sr^{-1}
	550 nm to 700 nm	T_1 to 10^{4}	$21 C_B$ J • cm^{-2} • sr^{-1}
	400 nm to 700 nm	10^{4} to 3×10^{4}	$2.1 C_B t \times 10^{-3}$ W • cm^{-2} • sr^{-1}
IR-A	700 nm to 1400 nm	10^{-9} to 10	$10 C_A \sqrt[3]{t}$ J • cm^{-2} • sr^{-1}
	700 nm to 1400 nm	10 to 10^{3}	$3.83 C_A (t/\sqrt[3]{t})$ J • cm^{-2} • sr^{-1}
	700 nm to 1400 nm	10^{3} to 3×10^{4}	$0.64 C_A$ W • cm^{-2} • sr^{-1}
IR-B & C	1.4 μm to 10^{3} μm	10^{-9} to 3×10^{4}	Same as Table 10

C_A, C_B, and T_1 are the same as in footnote to Table 10.

TABLE 12
Threshold Limit Value for Skin Exposure from a Laser Beam

Spectral Region	Wave Length	Exposure Time, (t) Seconds	TLV
UV	180 nm to 400 nm	10^{-9} to 3×10^{4}	Same as Table 10
Light & IR-A	400 nm to 1400 nm	10^{-9} to 10^{-7}	$2\, C_A \times 10^{-2}\ \mathrm{J} \bullet \mathrm{cm}^{-2}$
	" "	10^{-7} to 10	$1.1\, C_A \sqrt[4]{t}\ \mathrm{J} \bullet \mathrm{cm}^{-2}$
	" "	10 to 3×10^{4}	$0.2\, C_A\ \mathrm{W} \bullet \mathrm{cm}^{-2}$
IR-B & C	1.4 μm to 10^{3} μm	10^{-9} to 3×10^{4}	Same as Table 10

$C_A = 1.0$ for $\lambda = 400\text{-}700$ nm; see Figure 5 for $\lambda = 700$ to 1400 nm.

At wavelengths greater than 1400 nm, for beam cross-sectional areas exceeding 100 cm^2 the TLV for exposure durations exceeding 10 seconds is: $TLV = (10{,}000/A_s)\ mW/cm^2$
where A_s is the irradiated skin area for 100 to 1000 cm^2, and the TLV for irradiated skin areas exceeding 1000 cm^2 is 10 mW/cm^2 and for irradiated skin areas less than 100 cm^2 is 100 mW/cm^2.

NOISE

These Threshold Limit Values (TLVs) refer to sound pressure levels and durations of exposure that represent conditions under which it is believed that nearly all workers may be repeatedly exposed without adverse effect on their ability to hear and understand normal speech. Prior to 1979, the medical profession had defined hearing impairment as an average hearing threshold level in excess of 25 decibels (dB) (ANSI-S3.6-1969) at 500, 1000, and 2000 Hz, and the limits which are given have been established to prevent a hearing loss in excess of this level.[a] The values should be used as guides in the control of noise exposure and, due to individual susceptibility, should not be regarded as fine lines between safe and dangerous levels.

It should be recognized that the application of the TLV for noise will not protect all workers from the adverse effects of noise exposure. A hearing conservation program with audiometric testing is necessary when workers are exposed to noise at or above the TLV levels.

Continuous or Intermittent

The sound level shall be determined by a sound level meter, conforming as a minimum to the requirements of the American National Standard Specification for Sound Level Meters, S1.4 (1971) Type S2A, and set to use the A-weighted network with slow meter response.

These values apply to total duration of exposure per working day regardless of whether this is one continuous exposure or a number of short-term exposures and does include the impact and impulsive type of noise that contributes to the sound level meter reading at slow response.

When the daily noise exposure is composed of two or more periods of noise exposure of different levels, their combined effect should be considered, rather than the individual effect of each. If the sum of the following fractions:

$$\frac{C_1}{T_1} + \frac{C_2}{T_2} + \; \ldots \; \frac{C_n}{T_n}$$

exceeds unity, then, the mixed exposure should be considered to exceed the threshold limit value, C_1 indicates the total duration of exposure at a specific noise level, and T_1 indicates the total duration of exposure permitted at that level. All on-the-job noise exposures of 80 dBA or greater shall be used in the above calculations.

---

[a] In 1979 the American Academy of Ophthalmology and Otolaryngology (AAOO) included 3000 Hz in their hearing impairment formula.

TABLE 13
Threshold Limit Values for Noise

Duration per Day Hours	Sound Level dBA†
16	80
8	85
4	90
2	95
1	100
1/2	105
1/4	110
1/8	115*

† Sound level in decibels are measured on a sound level meter, conforming as a minimum to the requirements of the American National Standard Specification for Sound Level Meters, S1.4 (1971) Type S2A, and set to use the A-weighted network with slow meter response.

* No exposure to continuous or intermittent in excess of 115 dBA.

Impulsive or Impact

It is recommended that exposure to impulsive or impact noise shall not exceed the limits listed in Table 14 or taken from Figure 13. No exposures in excess of 140 decibels peak sound pressure level are permitted. Impulsive or impact noise is considered to be those variations in noise levels that involve maxima at intervals of greater than one per second. Where the intervals are less than one second, it should be considered continuous.

TABLE 14
Threshold Limit Values Impulsive or Impact Noise

Sound Level dB*	Permitted Number of Impulses or Impacts per day
140	100
130	1000
120	10,000

* Decibels peak sound pressure level; re 20 μPa.

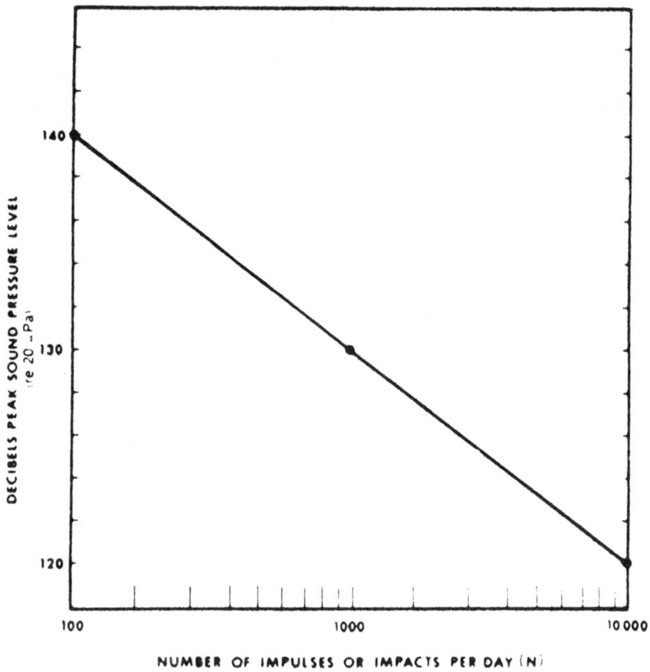

Figure 9—Threshold Limit Values for Impulse/Impact Noise.

RADIOFREQUENCY/MICROWAVE RADIATION

These Threshold Limit Values (TLVs) refer to radiofrequency (RF) and microwave radiation in the frequency range from 30 kHz to 300 GHz, and they represent conditions under which it is believed workers may be repeatedly exposed without adverse health effects. The TLVs shown in Table 15 are selected to limit the average whole body specific absorption rate (SAR) to 0.4 W/kg in any six-minute (0.1 hr) period for 3 MHz to 300 GHz; see Figure 10. Between 30 kHz and 3 MHz, the average whole body SAR is still limited to 0.4 W/kg, but the plateau at 100 mW/cm^2 was set to protect against shock and burn hazards.

Since it is usually impractical to measure the SAR, the TLVs are expressed in units that are measurable, viz, squares of the electric and magnetic field strength, averaged over any 0.1 hour period. This can be expressed in units of equivalent plane wave power density for convenience. The electric field strength (E) squared, magnetic field strength (H) squared, and power density

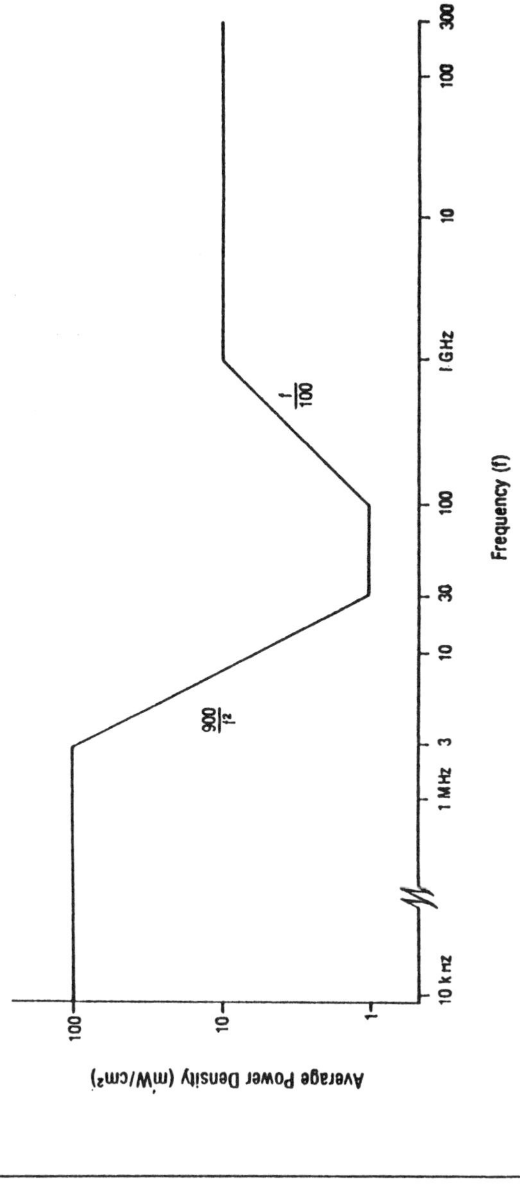

Figure 10 — Threshold Limit Values (TLV) for Radiofrequency/Microwave Radiation in the workplace (whole body SAR less than 0.4 W/kg).

(PD) values are shown in Table 15. For near field exposures PD cannot be measured directly, but equivalent plane wave power density can be calculated from the field strength measurement data as follows:

$$\text{PD in mW/cm}^2 = \frac{E^2}{3770}$$

where:

E^2 is in volts squared (V^2) per meter squared (m^2).

$$\text{PD in mW/cm}^2 = 37.7\ H^2$$

where:

H^2 is in amperes squared (A^2) per meter squared (m^2).

These values should be used as guides in the evaluation and control of exposure to radiofrequency/microwave radiation, and should not be regarded as a fine line between safe and dangerous levels.

Notes:

1. Needless exposure to all Radiofrequency Radiation (RFR) exposures should be avoided given the current state of knowledge on human effects, particularly nonthermal effects.
2. For fields consisting of a number of frequencies, the fraction of the protection guide incurred within each frequency level should be determined and the sum of all fractions should not exceed unity.
3. For pulsed and continuous wave fields, the power density is averaged over the six-minute period.
4. For partial body exposures at frequencies between 10 kHz and 1.0 GHz, the protection guides in Table 15 may be exceeded if the output power of a radiating device is 7 watts or less. For example, if a hand-held transmitter operating at 27 MHz has a maximum output of 5 watts, it would be excluded from any further field measurements.
5. The TLVs in Table 15 may be exceeded if the exposure conditions can be demonstrated to produce a SAR of less than 0.4 W/kg as averaged over the whole body and spatial peak SAR values less than 8.0 W/kg as averaged over any 1.0 gram of tissue. For example, for frequencies from 3 to 30 MHz, the equivalent power density can be increased by a factor of 10 up to a limit of 100 mW/cm^2, if it can be assured that exposed individuals are not in contact with the ground plate.
6. No measurement should be made within 5 cm of any object.
7. All exposures should be limited to a maximum (peak) electric field intensity of 100 kV/m.

TABLE 15
Radiofrequency/Microwave Threshold Limit Values

Frequency	Power Density (mW/cm^2)	Electric Field Strength Squared (V^2/m^2)	Magnetic Field Strength Squared (A^2/m^2)
10 KHz to 3 MHz	100	377,000	2.65
3 MHz to 30 MHz	$900/f^2$*	$3770 \times 900/f^2$	$900/(37.7 \times f^2)$
30 MHz to 100 MHz	1	3770	0.027
100 MHz to 1000 MHz	f/100	$3770 \times f/100$	$f/(37.7 \times 100)$
1 GHz to 300 GHz	10	37,700	0.265

*f = frequency in MHz.

‡ULTRAVIOLET RADIATION[A]

These Threshold Limit Values (TLVs) refer to ultraviolet radiation in the spectral region between 200 and 400 nm and represent conditions under which it is believed that nearly all workers may be repeatedly exposed without adverse effect. These values for exposure of the eye or the skin apply to ultraviolet radiation from arcs, gas and vapor discharges, fluorescent and incandescent sources, and solar radiation, but they do not apply to ultraviolet lasers.[A] These values do not apply to ultraviolet radiation exposure of photosensitive individuals or of individuals concomitantly exposed to photosensitizing agents.[(1)] These values should be used as guides in the control of exposure to continuous sources where the exposure duration shall not be less than 0.1 sec.

These values should be used as guides in the control of exposure to ultraviolet sources and should not be regarded as a fine line between safe and dangerous levels.

Recommended Values

The threshold limit values for occupational exposure to ultraviolet radiation incident upon skin or eye where irradiance values are known and exposure time is controlled are as follows:

1. For the near ultraviolet spectral region (320 to 400 nm), total irradiance incident upon the unprotected eye should not exceed 1.0 mW/cm^2 for periods greater than 10^3 seconds (approximately 16 minutes) and for exposure times less than 10^3 seconds should not exceed 1.0 J/cm^2.
2. For the actinic ultraviolet spectral region (200-315 nm), radiant exposure incident upon the unprotected skin or eye should not exceed the values given in Table 16 or Figure 11 within an 8-hour period.
3. To determine the effective irradiance of a broadband source weighted against the peak of the spectral effectiveness curve (270 nm), the following weighting formula should be used:

$$E_{eff} = \Sigma\, E_\lambda\, S_\lambda\, \Delta\lambda$$

where:

E_{eff} = effective irradiance relative to a monochromatic source at 270 nm in W/cm^2 (J/s/cm^2)

E_λ = spectral irradiance in W/cm^2/nm

S_λ = relative spectral effectiveness (unitless)

$\Delta\lambda$ = band width in nanometers

[A]See Laser TLVs.

‡See Notice of Intended Changes.

‡TABLE 16
Relative Spectral Effectiveness by Wavelength*

Wavelength (nm)	TLV (mJ/cm^2)	Relative Spectral Effectiveness S_λ
200	100	0.03
210	40	0.075
220	25	0.12
230	16	0.19
240	10	0.30
250	7.0	0.43
254	6.0	0.5
260	4.6	0.65
270	3.0	1.0
280	3.4	0.88
290	4.7	0.64
300	10	0.30
305	50	0.06
310	200	0.015
315	1000	0.003

*See Laser TLVs.
‡See Notice of Intended Changes.

TABLE 17
Permissible Ultraviolet Exposures

Duration of Exposure Per Day	Effective Irradiance, E_{eff} ($\mu W/cm^2$)
8 hrs	0.1
4 hr	0.2
2 hrs	0.4
1 hr	0.8
30 min	1.7
15 min	3.3
10 min	5
5 min	10
1 min	50
30 sec	100
10 sec	300
1 sec	3,000
0.5 sec	6,000
0.1 sec	30,000

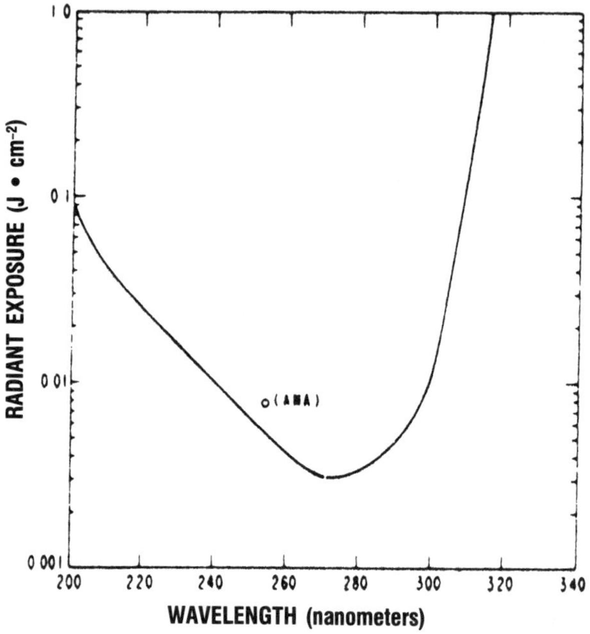

Figure 11—Threshold Limit Values for Ultraviolet Radiation.

4. Permissible exposure time in seconds for exposure to actinic ultraviolet radiation incident upon the unprotected skin or eye may be computed by dividing 0.003 J/cm^2 by E_{eff} in W/cm^2. The exposure time may also be determined using Table 17 which provides exposure times corresponding to effective irradiances in μW/cm^2.
5. All the preceding TLVs for ultraviolet energy apply to sources which subtend an angle less than 80°. Sources which subtend a greater angle need to be measured only over an angle of 80°.

Conditioned (tanned) individuals can tolerate skin exposure in excess of the TLV without erythemal effects. However, such conditioning may not protect persons against skin cancer.

Reference

1. *Sunlight and Man.* Fitzpatrick *et al*, Eds. Univ. of Tokyo Press, Tokyo, Japan (1974).

AIRBORNE UPPER SONIC AND ULTRASONIC ACOUSTIC RADIATION

These Threshold Limit Values (TLVs) refer to sound pressure levels that represent conditions under which it is believed that nearly all workers may be repeatedly exposed without adverse effect. The values listed in Table 19 should be used as guides in the control of noise exposure and, due to individual susceptibility, should not be regarded as fine lines between safe and dangerous levels. The levels for the third-octave bands centered below 20 kHz are below those which cause subjective effects. Those levels for 1/3 octaves above 20 kHz are for prevention of possible hearing losses from subharmonics of these frequencies.

TABLE 19
Permissible Airborne Upper Sonic and Ultrasound Acoustic Radiation Exposure Levels

Mid-Frequency of Third-Octave Band kHz	One-Third Octave—Band Level in dB re 20 μPa
10	80
12.5	80
16	80
20	105
25	110
31.5	115
40	115
50	115

Subjective annoyance may occur in some sensitive individuals at levels between 75 and 105 dB at 20 kHz 1/3 octave band and hearing protection or engineering controls may be needed to minimize or prevent the annoyance.

NOTICE OF INTENDED CHANGES (for 1989-90)

These physical agents, with their corresponding values, comprise those for which either a limit has been proposed for the first time, or for which a change in the "Adopted" listing has been proposed. In both cases, the proposed limits should be considered trial limits that will remain in the listing for a period of at least one year. If after one year no evidence comes to light that questions the appropriateness of the values herein, the values will be reconsidered for the the "Adopted" list.

ULTRAVIOLET RADIATION

The Committee voted to recommend a revision to the TLV for ultraviolet radiation (UV) in the UV-A band (315–400 nm) to more accurately reflect current biological data. This revision relaxes the limits for longer wavelengths and reduces the limits for the shorter UV-B wavelengths for lengthly exposures. The revision is as follows:

These Threshold Limit Values (TLVs) refer to ultraviolet radiation in the spectral region between 180 and 400 nm and represent conditions under which it is believed that nearly all workers may be repeatedly exposed without adverse effect. These values for exposure of the eye or the skin apply to ultraviolet radiation from arcs, gas and vapor discharges, fluorescent and incandescent sources, and solar radiation, but they do not apply to ultraviolet lasers (see the TLVs for Lasers). These values do not apply to ultraviolet radiation exposure of photosensitive individuals or of individuals concomitantly exposed to photosensitizing agents.(1) These exposures to the eye do not apply to aphakics. These values should be used as guides in the control of exposure to continuous sources where the exposure durations shall not be less than 0.1 sec.

These values should be used as guides in the control of exposure to ultraviolet sources and should not be regarded as a fine line between safe and dangerous levels.

Recommended Values

The threshold limit values for occupational exposure to ultraviolet radiation incident upon skin or eye where irradiance values are known and exposure time is controlled are as follows:

1. For the near ultraviolet spectral region (320 to 400 nm), total irradiance incident upon the unprotected skin or eye should

not exceed 1.0 mW/cm^2 for periods greater than 10^3 seconds (approximately 16 minutes) and for exposure times less than 10^3 seconds should not exceed 1.0 J/cm^2.

2. The ultraviolet radiant exposure incident upon the unprotected skin or eye should not exceed the values given in Table 16 or Figure 11 within an 8-hour period.
3. To determine the effective irradiance of a broadband source weighted against the peak of the spectral effectiveness curve (270 nm), the following weighting formula should be used:

$$E_{eff} = \Sigma E_\lambda S_\lambda \Delta\lambda$$

where:

E_{eff} = effective irradiance relative to a monochromatic source at 270 nm in W/cm^2 (J/s/cm^2)

E_λ = spectral irradiance in W/cm^2/nm

S_λ = relative spectral effectiveness (unitless)

$\Delta\lambda$ = band width in nanometers

4. For most white-light sources and all open arcs, the weighting of spectral irradiance between 200 and 315 nm should suffice to determine the effective irradiance. Only specialized UV sources designed to emit UV-A radiation would normally require spectral weighting from 315 to 400 nm.
5. Permissible exposure time in seconds for exposure to actinic ultraviolet radiation incident upon the unprotected skin or eye may be computed by dividing 0.003 J/cm^2 by E_{eff} in W/cm^2. The exposure time may also be determined using Table 17 which provides exposure times corresponding to effective irradiances in μW/cm^2.
6. All the preceding TLVs for ultraviolet energy apply to sources which subtend an angle less than 80°. Sources which subtend a greater angle need to be measured only over an angle of 80°.

Conditioned (tanned) individuals can tolerate skin exposure in excess of the TLV without erythemal effects. However, such conditioning may not protect persons against skin cancer.

Reference

1. *Sunlight and Man.* Fitzpatrick *et al,* Eds. Univ. of Tokyo Press, Tokyo, Japan (1974).

Table 17 and Figure 11 remain the same as they appear on pages 179 and 180, respectively.

TABLE 16
Utraviolet Radiation Exposure TLV and Spectral Weighting Function

Wavelength* (nm)	TLV (J/m²)	TLV (mJ/cm²)	Relative Spectral Effectiveness S_λ
180	2,500	250	0.012
190	1,600	160	0.019
200	1,000	100	0.030
205	590	59	0.051
210	400	40	0.075
215	320	32	0.095
220	250	25	0.120
225	200	20	0.150
230	160	16	0.190
235	130	13	0.240
240	100	10	0.300
245	83	8.3	0.360
250	70	7.0	0.430
254#	60	6.0	0.500
255	58	5.8	0.520
260	46	4.6	0.650
265	37	3.7	0.810
270	30	3.0	1.000
275	31	3.1	0.960
280#	34	3.4	0.880
285	39	3.9	0.770
290	47	4.7	0.640
295	56	5.6	0.540
297#	65	6.5	0.460
300	100	10	0.300
303#	250	25	0.190
305	500	50	0.060
308	1,200	120	0.026
310	2,000	200	0.015
313#	5,000	500	0.006
315	1.0×10^4	1.0×10^3	0.003
316	1.3×10^4	1.3×10^3	0.0024
317	1.5×10^4	1.5×10^3	0.0020
318	1.9×10^4	1.9×10^3	0.0016
319	2.5×10^4	2.5×10^3	0.0012

TABLE 16 (con't)
Utraviolet Radiation Exposure TLV and Spectral Weighting Function

Wavelength* (nm)	TLV (J/m^2)	TLV (mJ/cm^2)	Relative Spectral Effectiveness S_λ
320	2.9×10^4	2.9×10^3	0.0010
322	4.5×10^4	4.5×10^3	0.00067
323	5.6×10^4	5.6×10^3	0.00054
325	6.0×10^4	6.0×10^3	0.00050
328	6.8×10^4	6.8×10^3	0.00044
330	7.3×10^4	7.3×10^3	0.00041
333	8.1×10^4	8.1×10^3	0.00037
335	8.8×10^4	8.8×10^3	0.00034
340	1.1×10^5	1.1×10^4	0.00028
345	1.3×10^5	1.3×10^4	0.00024
350	1.5×10^5	1.5×10^4	0.00020
355	1.9×10^5	1.9×10^4	0.00016
360	2.3×10^5	2.3×10^4	0.00013
365#	2.7×10^5	2.7×10^4	0.00011
370	3.2×10^5	3.2×10^4	0.000093
375	3.9×10^5	3.9×10^4	0.000077
380	4.7×10^5	4.7×10^4	0.000064
385	5.7×10^5	5.7×10^4	0.000053
390	6.8×10^5	6.8×10^4	0.000044
395	8.3×10^5	8.3×10^4	0.000036
400	1.0×10^6	1.0×10^5	0.000030

* Wavelengths chosen are representative; other values should be interpolated at intermediate wavelengths.

Emission lines of a mercury discharge spectrum.

NOTICE OF INTENT TO ESTABLISH THRESHOLD LIMIT VALUES

LIGHT AND NEAR-INFRARED RADIATION

These Threshold Limit Values (TLVs) refer to visible and near-infrared radiation in the wavelength range of 400 nm to 1400 nm and represent conditions under which it is believed that nearly all workers may be exposed without adverse effect. These values should be used as guides in the control of exposure to light and should not be regarded as a fine line between safe and dangerous levels.

Recommended Values

The Threshold Limit Value for occupational exposure to broad-band light and near-infrared radiation for the eye apply to exposure in any 8-hour workday and require knowledge of the spectral radiance (L_λ) and total irradiance (E) of the source as measured at the position(s) of the eye of the worker. Such detailed spectral data of a white light source are generally only required if the luminance of the source exceeds 1 cd cm^{-2}. At luminances less than this value, the TLV would not be exceeded.

The TLVs are:

1. To protect against retinal thermal injury, the spectral radiance of the lamp weighted against the function R (Table 18) should not exceed:

$$\sum_{400}^{1400} L_\lambda \bullet R_\lambda \bullet \Delta\lambda \leqslant 1/\alpha t^{1/2} \qquad (1)^*$$

where L_λ is in W cm^{-2} sr^{-1} nm^{-1} and t is the viewing duration (or pulse duration if the lamp is pulsed) limited to 1 μs to 10 s, and α is the angular subtense of the source in radians. If the lamp is oblong, α refers to the longest dimension that can be viewed. For instance, at a viewing distance r = 100 cm from a tubular lamp of length l = 50 cm, the viewing angle is:

$$\alpha = l/r = 50/100 = 0.5 \text{ rad} \qquad (2)$$

* Formulae (1) and (7) are empirical and are not, strictly speaking, dimensionally correct. To make the formulae dimensionally correct, one would have to insert a dimensional correction factor *k* in the right hand numerator in each formula. For formula (1) this would be k_1 = 1 W • rad • $s^{1/2}$/(cm^2 • sr), and for formula (7) K_2 = 1 W • rad/(cm^2 • sr).

2. To protect against retinal photochemical injury from chronic blue-light exposure, the integrated spectral radiance of a light source weighted against the blue-light hazard function B_λ (Table 18) should not exceed:

$$\sum_{400}^{700} L_\lambda \bullet t \bullet B_\lambda \bullet \Delta\lambda \leqslant 100 \text{ J} \bullet \text{cm}^{-2} \bullet \text{sr}^{-1} \ (t \leqslant 10^4 \text{ s}) \quad (3a)$$

$$\sum_{400}^{700} L_\lambda \bullet B_\lambda \bullet \Delta\lambda \leqslant 10^{-2} \text{ W} \bullet \text{cm}^{-2} \bullet \text{sr}^{-1} \ (t > 10^4 \text{ s}) \quad (3b)$$

The weighted product of L_λ and B_λ is termed L (blue). For a source radiance L weighted against the blue-light hazard function (L [blue]) which exceeds 10 mW cm^{-2} sr^{-1} in the blue spectral region, the permissible exposure duration t_{max} in seconds is simply:

$$t_{max} = 100 \text{ J cm}^{-2} \text{ sr}^{-1}/\text{L (blue)} \quad (4)$$

The latter limits are greater than the maximum permissible exposure limits for 440 nm laser radiation (*see* Laser TLV) because a 2–3 mm pupil is assumed rather than a 7-mm pupil for the Laser TLV. For a light source subtending an angle α less than 11 mrd (0.011 radian), the above limits are relaxed such that the spectral irradiance weighted against the blue-light hazard function B_λ should not exceed E (blue).

$$\sum_{400}^{700} E_\lambda \bullet t \bullet B_\lambda \bullet \Delta\lambda \leqslant 10 \text{ mJ} \bullet \text{cm}^{-2} \ (t \leqslant 10^4 \text{ s}) \quad (5a)$$

$$\sum_{400}^{700} E_\lambda \bullet B_\lambda \bullet \Delta\lambda \leqslant 1 \ \mu\text{W} \bullet \text{cm}^{-2} \ (t \geqslant 10^4 \text{ s}) \quad (5b)$$

For a source where the blue light weighted irradiance E (blue) exceeds 1 μW • cm^{-2}, the maximum permissible exposure duration t_{max} in seconds is:

$$t_{max} = 10 \text{ mJ} \bullet \text{cm}^{-2}/\text{E (blue)} \quad (6)$$

For the worker having a lens removed (cataract surgery), there will be an increased blue-light hazard. And unless a UV-absorbing intra-ocular lens has been surgically inserted into the eye, an adjusted B_λ function should be used in equations 3a, 3b, 5a, and 5b and the function summed from 300 to 700 nm. This alternative B_λ function is termed the Aphakic Hazard Function, A_λ.

3. *Infrared radiation:* To avoid possible delayed effects upon the lens of the eye (cataractogenesis), the infrared radiation (λ = 770 nm) should be limited to 10 mW cm^{-2}. For an infrared heat lamp or any near-infrared source where a strong visual stimulus is absent, the near infrared (770-1400 nm) radiance as viewed by the eye should be limited to:

$$\sum_{770}^{1400} L_\lambda \bullet \Delta\lambda \leqslant 0.6/\alpha \qquad (7)^*$$

for extended duration viewing conditions. This limit is based upon a 7-mm pupil diameter.

TABLE 18
Spectral Weighting Functions for Assessing Retinal Hazards from Broad-Band Optical Sources

Wavelength (nm)	Blue-Light Hazard Function B_λ	Aphake Photic Hazard Function A_λ	Burn Hazard Function R_λ
305	—	6.00	—
310	—	6.00	—
315	—	6.00	—
320	—	6.00	—
325	—	6.00	—
330	—	6.00	—
335	—	6.00	—
340	—	5.88	—
345	—	5.71	—
350	—	5.46	—
355	—	5.22	—
360	—	4.62	—
365	—	4.29	—
370	—	3.75	—
375	—	3.56	—
380	—	3.19	—

TABLE 18 (con't)
Spectral Weighting Functions for Assessing Retinal Hazards from Broad-Band Optical Sources

Wavelength (nm)	Blue-Light Hazard Function B_λ	Aphake Photic Hazard Function A_λ	Burn Hazard Function R_λ
385	—	2.31	—
390	—	1.88	—
395	—	1.58	—
400	0.10	1.43	1.0
405	0.20	1.30	2.0
410	0.40	1.25	4.0
415	0.80	1.20	8.0
420	0.90	1.15	9.0
425	0.95	1.11	9.5
430	0.98	1.07	9.8
435	1.0	1.03	10.0
440	1.0	1.0	10.0
445	0.97	0.97	9.7
450	0.94	0.94	9.4
455	0.90	0.90	9.0
460	0.80	0.80	8.0
465	0.70	0.70	7.0
470	0.62	0.62	6.2
475	0.55	0.55	5.5
480	0.45	0.45	4.5
485	0.40	0.40	4.0
490	0.22	0.22	2.2
495	0.16	0.16	1.6
500-600	$10^{[(450-\lambda)/50]}$	$10^{[(450-\lambda)/50]}$	1.0
600-700	0.001	0.001	1.0
700-1049	NA*	NA*	$10^{[(700-\lambda)/500]}$
1050-1400	NA	NA	0.2

* NA = Not applicable.

STATIC MAGNETIC FIELDS

These Threshold Limit Values (TLVs) refer to static magnetic flux densities to which it is believed that nearly all workers may be repeatedly exposed day after day without adverse effects. These values should be used as guides in the control of exposure to static magnetic fields and should not be regarded as a fine line between safe and dangerous levels.

Routine occupational exposures should not exceed 60 milliTeslas (mT)—equivalent to 600 Gauss—whole body or 600 mT (6000 Gauss) to the extremities on a daily, time-weighted average basis. A flux density of 2 Teslas is recommended as a ceiling value. Safety hazards may exist from the mechanical forces exerted by the magnetic field upon ferromagnetic tools and medical implants. Workers having implanted cardiac pacemakers should not be exposed above 1.0 mT (10 Gauss). Perceptible or adverse effects may also be produced at higher flux densities resulting from forces upon other implanted ferromagnetic medical devices, e.g., suture staples, aneurism clips, prostheses, etc.

EXTREMELY LOW FREQUENCY (ELF) ELECTRIC FIELDS

These Threshold Limit Values (TLVs) refer to field strengths of Extremely Low Frequency (ELF) Electric Fields (up to 30 kHz) that represent conditions under which it is believed that most workers may be exposed repeatedly without permanent biological effects from exposure to a maximum unprotected workplace field strength level for ELF radiation. The value should be used as a guide in the control of exposure and, due to individual susceptibility, should not be regarded as a fine line between safe and dangerous levels.

Routine occupational exposures should not exceed 25 kV/m.

Notes:

1. The potential hazards of high voltage 60 Hz electric fields are recognized.
2. Field intensities of 5 kV/m can be sensed by almost everyone under the proper conditions.
3. Field intensities greater than 10 kV/m in the workplace can produce a wide range of safety hazards associated with shock, spark discharges, and other startle reactions.

4. At frequencies below 30 kHz, ungrounded objects such as vehicles, fences, etc., can strongly couple to RF fields. For field strengths near the TLV, shock and burn hazards can exist. Care should be taken to eliminate ungrounded objects, to ground such objects, or use insulated gloves when ungrounded objects must be handled.
5. Prudence dictates the use of protective devices (suits, gloves, insulation, etc.) in all fields exceeding 15 kV/m.

PHYSICAL AGENTS UNDER STUDY

The Physical Agents TLV Committee of ACGIH has examined the current literature and has not found sufficient information to propose a TLV. However, these agents will remain under study during the coming year to examine new evidence indicating the need and feasibility for establishing a proposed TLV. Comments and suggestions, accompanied by substantive documentation are solicited and should be forwarded to the Executive Secretary, ACGIH. Documentation summarizing the current status of the biological effects literature is available on those agents preceded by an asterisk (*).

1. *Low Frequency Magnetic Fields.*
2. *Laser Radiation.* Specifically laser exposures of less than one (1) nanosecond.
3. *Vibration.* Whole-body.
4. *Pressure Variations.*
5. *Electromagnetic Pulses.*
6. *Radiofrequency Radiation* (RFR). Peak power limits.
7. *Static Electric Fields.*

Appendix E

Recommended Reference Materials

Accident Prevention Manual for Industrial Operations: Administration and Program, 8th ed. (Chicago, IL: National Safety Council).

Advances in Air Sampling, American Conference of Governmental Industrial Hygienists (Chelsea, MI: Lewis Publishers, Inc., 1988).

Air Sampling Instruments for Evaluation of Atmospheric Contaminants, 6th ed. (Cincinnati, OH: American Conference of Governmental Industrial Hygienists).

"The Basic Science of Poisons," in *Casarett and Doull's Toxicology,* current ed. J. Doull, C. D. Klaassen, and M. O. Amdur, Eds. (New York: MacMillan Publishing Co., Inc.).

Chemical Hazards in the Workplace, R. M. Scott (Chelsea, MI: Lewis Publishers, Inc., 1989).

Clinical Toxicology of Commercial Products, 5th ed. (Baltimore: Williams and Wilkins, 1984).

Code of Federal Regulations. Title 29, Section 1910. Washington, DC.

Condensed Chemical Dictionary, 10th ed. (New York: Van Nostrand Reinhold Co., 1984).

Dangerous Properties of Industrial Materials, current ed., N. I. Sax (New York: Van Nostrand Reinhold Co., 1984).

Dorland's Illustrated Medical Dictionary, current ed. (Philadelphia: W. B. Saunders Co.).

Effects of Exposure to Toxic Gases: First Aid and Medical Treatment,

2nd ed. (Lyndhurst, NJ: Matheson, Division of Searle Medical Products, 1982).

Encyclopedia of Occupational Health and Safety, 3rd revised ed. Vols. 1 and 2. (Washington, DC: International Labor Office).

Engineering Field Reference Manual (Akron, OH: American Industrial Hygiene Association).

Ergonomic Design for People at Work, Vols. 1 and 2. Eastman Kodak (Florence, KY: Wadsworth, Inc.).

Ergonomic Interventions to Prevent Musculoskeletal Injuries in Industry, American Conference of Governmental Industrial Hygienists (Chelsea, MI: Lewis Publishers, Inc., 1987).

Fire Protection Handbook, 15th ed. NFPA No. FPH 1581 (Quincy, MA: National Fire Protection Association International).

Fundamentals of Industrial Hygiene, 3rd ed. (Chicago: National Safety Council, 1983).

Guide to Industrial Respiratory Protection, J. A. Pritchard (Washington, DC: National Technical Information Service).

Handbook of Chemistry and Physics, 65th ed. R. C. Weast, M. J. Astle, and W. H. Beyer, Eds. (Boca Raton, FL: CRC Press, Inc.).

Handbook of Hazardous Materials Fire Safety Health, (Schaumburg, IL: Alliance of American Insurers).

Handbook of Organic Industrial Solvents, 6th ed. (Schaumburg, IL: Alliance of American Insurers).

Handbook of Reactive Chemical Hazards, 3rd ed., L. Bretherick (Stoneham, MA: Butterworth Publishers, Inc.).

Handbook of Toxic and Hazardous Chemicals and Carcinogens, current ed. (Park Ridge, NJ: Noyes Data Corporation, 1985).

Hazard Assessment and Control Technology In Semiconductor Manufacturing, American Conference of Governmental Industrial Hygienists (Chelsea, MI: Lewis Publishers, Inc., 1989).

Health and Safety in Small Industry: A Practical Guide for Managers, University of Medicine and Dentistry of New Jersey–Robert Wood Johnson Medical School (Lewis Publishers, Inc., 1989).

Human Factors Design Handbook, W. E. Woodson (New York: McGraw-Hill Book Company).

Indoor Air and Human Health, R. B. Gammage and S. V. Kaye, Eds. (Chelsea, MI: Lewis Publishers, Inc., 1985).

The Industrial Environment—Its Evaluation and Control (Cincinnati, OH: National Institute of Occupational Safety and Health).

Industrial Hazard and Safety Handbook, R. W. King and J. Magid (Stoneham, MA: Butterworth Publishers, Inc.).

Industrial Hygiene Engineering, 2nd ed., J. T. Talby (Park Ridge, NJ: Noyes Data Corporation).

Industrial Medicine Desk Reference, D. F. Tyer and K. A. Anderson (New York: Chapman and Hall).

Industrial Noise and Hearing Conservation (Chicago: National Safety Council).

Industrial Noise Control Handbook, P. N. Cheremisinoff and P. P. Cheremisinoff (Ann Arbor, MI: Ann Arbor Science Publishers).

Industrial Noise Manual, current ed. (Akron, OH: American Industrial Hygiene Association).

Industrial Solvents Handbook, 2nd ed. (Park Ridge, NJ: Noyes Data Corporation).

Industrial Toxicology, 4th ed., A. J. Finkel (Boston: John Wright, PSG, Inc., 1983).

Industrial Ventilation: A Manual of Recommended Practice, current ed. (Lansing, MI: ACGIH Committee on Industrial Ventilation).

Industrial Ventilation—A Self Study Companion to the ACGIH Ventilation Manual, J. D. Burton (Salt Lake City, UT: IVE, Inc.).

Lowrys' Handbook of Right-to-Know and Emergency Planning, G. G. Lowry and R. C. Lowry (Chelsea, MI: Lewis Publishers, Inc., 1988).

Manual of Analytical Methods, 3rd ed. NIOSH (Washington, DC: National Technical Information Service).

Matheson Guide to Safe Handling of Compressed Gases (Lyndhurst, NJ: Matheson, Division of Searle Medical Products, 1982).

The Merck Index, 10th ed. (Rahway, NJ: Merck & Co., Inc.).

Microcomputer Applications in Occupational Safety and Health, American Conference of Governmental Industrial Hygienists (Chelsea, MI: Lewis Publishers, Inc., 1987).

Occupational Biomechanics, D. B. Chaffin and G. Anderson (New York: Wiley-Interscience).

Occupational Health Guidelines for Chemical Hazards. NIOSH/OSHA (Washington, DC: U.S. Government Printing Office).

Occupational Health Practice, 2nd ed., R. S. F. Schilling (Stoneham, MA: Butterworth Publishers, Inc.).

Patty's Industrial Hygiene and Toxicology, 2nd ed., Vols. 1, 2A, 3A, 3B. G. D. Clayton and F. E. Clayton, Eds. (New York: John Wiley and Sons, Inc.).

Patty's Industrial Hygiene and Toxicology, 3rd ed., Vols. 1-5. G. D. Clayton and F. E. Clayton, Eds. (New York: John Wiley and Sons, Inc., 1978).

Pocket Guide to Chemical Hazards, F. W. Mackison et al., Eds. NIOSH/OSHA (Washington, DC: U.S. Government Printing Office Pub. No. 78-210).

Respiratory Protection, American Industrial Hygiene Association Monograph Series (Akron, OH: American Industrial Hygiene Association).

Threshold Limit Values and Biological Exposure Indices, current ed. (Cincinnati, OH: American Conference of Governmental Industrial Hygienists).

Toxicological Chemistry, S. E. Manahan (Chelsea, MI: Lewis Publishers, Inc., 1988).

Toxicology, M. A. Kamrin (Chelsea, MI: Lewis Publishers, Inc., 1987).

Appendix F

Glossary

ABIH: American Board of Industrial Hygiene. A professional industrial hygiene organization. Major activity is certification and recertification of industrial hygienists.

ACGIH: American Conference of Governmental Industrial Hygienists. A professional organization that provides consensus standards for occupational exposure to chemical and physical agents.

Acute poisoning: This refers to a condition produced usually by a single exposure or a few repeated exposures.

AIHA: American Industrial Hygiene Association. A professional industrial hygiene organization. Major activities are laboratory accreditation and journal publication.

Airborne contaminant: A chemical element, compound, or mixture that is airborne in the form of a gas, vapor, or aerosol (dust, mist, fume, smoke).

Air cleaner: Device designed to remove airborne contaminants such as dust, gases, vapors, mists, and smoke. Common devices include filters, cyclones, scrubbers, charcoal beds, and electrostatic precipitators.

Anemia: A condition in which the blood is deficient either in quantity or quality.

Anesthesia: The entire or partial loss of feeling or sensation with paralysis.

ANSI: American National Standards Institute.

Asbestos: A hydrated magnesium silicate in fibrous form.

Audiogram: Results of a hearing test; a record of hearing level tested at several different sound frequencies.

Blank: Unexposed sample analyzed with test samples to verify accuracy of sampling techniques and analytical methods.

BOHS: British Occupational Hygiene Society. Principle function is to promote occupational hygiene.

Breathing zone sample: An air sample collected in the employee's breathing zone to assess individual exposure.

Bronchitis: Inflammation of the air passages, large and small, in the lungs.

BSC: British Safety Council. A nonprofit organization registered as a charity.

BSI: British Standards Institute.

Canadian Center for Occupational Health and Safety: Founded by an act of the Canadian parliament in 1978. Its purpose is "to promote the fundamental right of Canadians to a healthy and safe working environment."

Capture velocity: The velocity required to capture or entrain a contaminant into the exhaust ventilation system.

Carcinogen: Any cancer-producing substance.

Carcinoma: A malignant growth made up of cells tending to infiltrate surrounding tissues.

Central nervous system: The brain and the spinal cord.

Chemical agent: A chemical element, compound, or mixture which may cause adverse health effects through exposure. Exposure may result from absorption through skin or mucous membranes, inhalation of airborne forms (dusts, mists, fumes, vapors), or ingestion.

Chronic poisoning: A type of disease or illness occurring as the result of exposure to small amounts of poisonous materials usually over a prolonged period of months or years. Chronic effects may sometimes occur following short-term exposure.

Combustible: Anything that will burn.

Conjunctivitis: Inflammation of the delicate membrane lining the lids and covering the eyeballs.

Corneal damage: Damage to the transparent part of the coating of the eyeball that covers the iris and pupil and admits light.

CPR: Cardiopulmonary resuscitation.

Cyanosis: Blueness of the skin caused by lack of oxygen in the blood.

Decibel (dB): A unit for measurement of sound levels. For noise exposure measurement, instruments are used that resemble the human ear in sensitivity to different noise frequencies. These instruments provide an "A-weighted" sound level in units called dBA.

Dermatitis: An inflammatory skin condition.

DIN: Deutches Institut für Normung e.V., German catalog of technical rules.

Dusts: Solid particles of matter produced by grinding, crushing, etc. Size may range from 0.5 to 100 μm (1 μm = 1/25000 inch) with most (over 90%) of airborne dust in the 0.5- to 5-μm range.

Edema (lung): Fluid in the lung.

Engineering controls: Methods of reducing exposure by controlling at the source or reducing the quantity of contaminant or agent released. Examples include design items or modifications in equipment, ventilation, processes, or tools that reduce potential for employee exposure.

Explosive limits: The concentration range, usually given in percent by volume, of a combustible gas or vapor that must be present in air before ignition can take place. Limits may vary from 1 to 50% for various materials. Normally, only the lower explosive limit is important.

Flammable liquid: Any liquid that will burn. The lower the flash point, the greater the flammability.

Flash point: The lowest temperature of a flammable liquid at which enough vapors are given off to form a flammable mixture near the liquid surface. A flame, spark, or other ignition source near the liquid surface will cause ignition.

Fumes: Solid particles of matter formed by condensation of vapors. They are usually produced by heating metal or other solids. Size usually ranges from 0.01 to 0.5 μ.

Gas: Defined as a material above its critical temperature. The name is usually applied to any material which is in the gaseous state at normal temperature and pressure.

Gastrointestinal: Pertaining to the stomach and intestine.

Health and Safety Act: The United Kingdom's equivalent of OSHA, administered by the Health and Safety Commission.

Heat stress: Physiological response to thermal load presented by combinations of convective heat, radiant heat, and metabolic activity. Effects may range from discomfort to heat exhaustion. May be controlled by modification of the following factors: work rate and schedule, source of radiant heat, employee clothing, air temperature, and relative humidity.

Ionizing radiation: Radiation with sufficient energy to cause ionization. There are five major types: alpha, beta, X-ray, gamma, and neutron.

ISO: International Standards Organization.

Job hazard analysis: An analysis consisting of studying each element of the job, recording observations, and identifying hazards involved in the process operations, employee work practices, or other job aspects. Procedures to be followed to control or correct each hazard can then be defined and implemented.

Lachrymation: Secretion and discharge of tears.

Laryngitis: Inflammation of the organ of voice, with dryness and soreness of the throat, hoarseness and cough, and difficulty in swallowing.

Local exhaust ventilation: A system designed to remove a contaminant near its source of generation. Characteristically includes a hood designed to effectively capture the contaminant.

Makeup air: Air required to replace the quantity removed or lost by the building ventilation system. Ensures proper balancing and optimal operation of the system.

MAK values (German): The maximum permissible concentration of a chemical compound in air in a working area that, according to current knowledge, generally does not impair the health of an employee or cause undue annoyance.

Maximum allowable concentration: The highest concentration to which a person may be continually exposed (eight hours a day, five days a week) without danger of damage to health. The values are not exact and are subject to revision as additional information is obtained. No maximum allowable concentration has been established for mixtures of materials, but when a person is exposed simultaneously to two or more materials this fact must be considered in establishing safe limits. Short exposures to concentrations higher than the maximum allowable concentration may not be harmful, depending on the length of time, the concentration, and the nature of the material.

While maximum allowable concentrations may vary among countries, the information referenced here is recommended by the American Conference of Governmental Industrial Hygienists. The ACGIH standards are reproduced by many countries in their national standards and/or serve as a base for their standards.

Milligrams per cubic meter (mg/m^3): Refers to the concentration of a solid or liquid suspended in air and indicates the weight present in a given volume of air. One ounce equals 28,000 milligrams; 35 cubic feet is approximately one cubic meter.

Mists: Liquid particles produced by such processes as spraying, atomization, or mechanical agitation.

MSDS: Material Safety Data Sheet. A chemical, product, or material fact sheet supplied by a manufacturer or distributor. The sheet gives information on contents, hazardous ingredients, and health and safety hazards, as well as medical first aid information.

MSHA: (U.S.) Mine Safety and Health Administration. Regulatory agency responsible for occupational safety and health in mining industries.

Mucous membrane: A thin layer of tissue lining the canals and cavities of the body that communicate with external air. Corrosion of mucous membranes is caused by a substance with a caustic and locally destructive effect on this tissue.

Mutagen: A chemical or physical agent that induces genetic mutations.

Narcotic: Producing sleep or stupor.

NBS: (U.S.) National Bureau of Standards.

Neuromuscular system: The nerves of the body and the muscles that they control.

NIOSH: National Institute of Occupational Safety and Health. U.S. federal agency responsible for research and guidance in areas of occupational safety and health.

Noise: Unwanted sound. The disturbing effects of noise depend on both intensity and frequency. The human ear is more sensitive to high frequencies.

Non-ionizing radiation: Radiation in the region of the electromagnetic spectrum not capable of producing ionization. Includes ultraviolet and visible light, infrared, microwave, and radio frequencies.

Nuisance: A condition that may be objectionable, annoying, or disagreeable, but having no known toxic or harmful effect.

Occupational hygienist: A professional in recognition, evaluation, and control of occupational health hazards. A certified industrial hygienist is an individual with appropriate training and experience who has been reviewed, through examination, by the ABIH and participates in its certification maintenance program.

Odor threshold: This refers to the lowest concentration of odor detectable by the average person. Because of individual differences in the sense of smell, this figure serves only as a very rough guide.

OSHA: Occupational Safety and Health Administration. U.S. regulatory agency responsible for promulgation and enforcement of occupational health and safety regulations.

Parts per million (ppm): This refers to the concentration of a gas or vapor in air, and indicates the volume present per million volumes of air.

Physical agents: A form of energy that can produce adverse health effects through exposure. Included are mechanical energy, noise, vibration, electromagnetic energy (visible light, ultraviolet light, infrared light, and microwave and radio frequency energies), and energies capable of ionization (alpha, beta, or gamma radiation, neutrons, and X-rays).

Polymerization: The uniting of two or more molecules of a substance.

Pulmonary: Pertaining to the lungs.

Pulmonary hemorrhage: Hemorrhage, or bleeding, from the lungs.

Respiratory protection: Devices used to reduce inhalation of airborne contaminants. May be air-purifying or supply air type. Air purifiers may contain a mechanical filter (for dusts, particulates, aerosols, fumes, and fibers), charcoal canister (for vapors), or other absorbent material.

RWP: Radiation Work Permit. An administrative permit system restricting access to, and activities in, a radiation area.

SCBA: Self-contained breathing apparatus. A type of respiratory protection in which clear air or oxygen is supplied directly to user.

Sensitizing agents: An agent that may not produce any reaction on the skin on initial contact up to 24 hours, but that will initiate a body process so that a second application several weeks later, and future applications, will produce a definite reaction. Most sensitizing agents are organic chemicals, but almost any material may act in such a way in susceptible individuals.

Skin absorption: The taking up of fluids, vapors, or other substances by the skin, from which they are eventually transferred to the bloodstream.

Skin irritant: A substance (usually called primary skin irritant) that, when in contact with the skin for 24 hours or less, will produce inflammation. All acids and alkalis are skin irritants.

Smoke: Usually refers to the products of combustion that are visible because of unburned carbon.

Split samples: Division of sample into two identical samples with the purpose of determining laboratory quality through the detection of analytical variations/error.

Teratogen: An agent or factor that causes physical defects in the developing embryo.

TLV-STEL: ACGIH Threshold Limit Value–Short-Term Exposure Limit. The maximum short-term (15-minute) allowable exposure concentration.

TLV-TWA: ACGIH Threshold Limit Value–Time-Weighted Average. The time-weighted average concentration of airborne contaminants for a normal eight-hour workday (40-hour workweek) to which nearly all workers may be repeatedly exposed without adverse effect. A maximum permissible eight-hour average exposure concentration.

Toxicity: Refers to systemic poisoning action. While almost anything may be toxic in sufficient quantity, it is used here to refer to the condition that would be produced in normal industrial conditions.

Toxicologist: A person specially trained to determine the nature of the toxic effects of substances and the probability of their occurrence under various conditions.

Toxicology: The study of the adverse effects of substances on living organisms.

Trauma: A wound or injury.

TWA: Time-weighted average. The average concentration or dose determined by personal sampling during all or a portion of an individual's eight-hour workday. For comparison with the TLV-TWA, the average concentration over a full eight-hour work day should be determined.

Ulceration: The loss of substance on a cutaneous or mucous surface causing gradual disintegration and necrosis (death) of the tissues.

Vapor: The gaseous state of a material that is solid or liquid at normal temperature.

Ventilation: A primary method of exposure control. May be designed as a general or local system. General ventilation reduces contaminant concentration by dilution (continually adding "fresh air"). Local exhaust ventilation captures and removes contaminants from the employee's breathing zone at the point of generation.

Vertigo: Giddiness or dizziness, which can be caused by some poison in the blood.

Wet Bulb Globe Temperature (WBGT): A measurement used to evaluate heat stress. Includes the following components: natural wet bulb temperature, dry bulb air temperature, and globe temperature. With knowledge of metabolic load, WBGT can be used to determine allowable work/rest cycles. (See Appendix D.)